HOWS AND KNOTTS

HOWS AND KNOTTS

A Guide to Lakeland Views

GUY RICHARDSON

First published in 2019 by Redshank Books

Redshank Books is an imprint of Libri Publishing.

ISBN 978-1-912969-05-0

A CIP catalogue record for this book is available from The British Library

Cover and book design by Carnegie Publishing

The support of the Lake District National Park Authority in the publication of this guide is gratefully acknowledged

The author is making regular donations to the Lake District Foundation following the publication of this guide

www. lakedistrictfoundation.org

Lake District National Park

Printed in the UK by Short Run Press Ltd

Libri Publishing
Brunel House
Volunteer Way
Faringdon
Oxfordshire
SN7 7YR

Tel: +44 (0)845 873 3837

www.libripublishing.co.uk

Maps on pages 21, 52, 62, 68, 74, 80, 86, 92, 98, 104, 110, 116, 122, 128, 134, 140, 146, 152, 158, 164, 170, 176, 182, 188 © Crown Copyright and Database Rights 2019 Ordnance Survey Licence 100061047

Front cover photo: Looking north-west from Dod Knott towards the elevated prehistoric landscape of Eskdale Moor in the middle distance
Frontispiece photo: Looking north-west over Crummock Water from Rannerdale Knotts

"We see nothing truly until we understand it"
John Constable (1776–1837), English landscape painter

Contents

Introduction

What do you see when you look at a view?

Imagine you've just finished a walk over the Langdale Pikes. Your day started with that famous view of the Pikes as you made your way up Langdale and the sight of those shapely and lofty summits inspired you to put on your boots and make the effort. You climbed slowly, passing through craggy fell country while enjoying the widening views. You clambered over the summits in turn, choosing to finish on Harrison Stickle, the highest hereabouts, where you soaked up the vista and shared with your companion the ritual of naming the distant peaks. Then you started your descent, looking down at changing valley scenes as your path twisted, turned and dropped ever lower until you were down on the valley floor, weary and enjoying that cup of tea or pint of beer. It's been a great day with good views, good companionship and good memories. What more could there possibly be?

A classic view of the Langdale Pikes

Now let's imagine making a minor diversion during your descent. Coming down under Harrison Stickle you have carefully traversed the precipitous ground above the chasm of Dungeon Ghyll and a broad shoulder of fellside has opened out. Things are a little easier now but as you walk on down you become aware that the steeper ground some way either side of you is gradually closing in. The further you walk, the closer the steep ground becomes until you see the path diving off right downwards. But you don't follow the path, you keep on going. You clamber over and along a rocky promontory to its far end where, like an actor reaching the extremity of a thrust stage, you suddenly feel very exposed. This airy spot is Pike Howe.

Pike Howe is in the dead centre of the classic view of the Langdale Pikes from lower down the valley. As you gaze out from your perch you can imagine scores of eyes looking in your direction. Hundreds of thousands of people have stared at, photographed and sketched Pike Howe without realising it. So, as you look down the valley towards those camera lenses, what do you make of this reverse view? How would you describe it? You are familiar with the view looking towards the Pikes, which you might describe as inspiring or uplifting. You are familiar with the vistas from the summits, which you might describe as exhilarating or plain awesome. So which word to describe this view from Pike Howe? Adjectives like "beautiful" or even that old fashioned word "picturesque" might come to mind. Less likely, perhaps, is the word "interesting". But *interesting* it really is and even more than that – the view is a real story teller. Every feature you see, natural and man-created, is where it is and looks as it does because of its place in the evolutionary story of this particular view.

The Lake District is now a designated UNESCO World Heritage Site. It has received this accolade because it is considered to be a "cultural landscape" of world significance. UNESCO defines cultural landscapes as representing "the combined works of nature and of man". In this context "cultural" is meant in its broadest sense: how the skills, brawn, intellect and perseverance of everyone in a society have collectively shaped a landscape. The view from Pike Howe is a superb example of a cultural landscape. The volcanic rocks around you, the bends in the valley, the standing walls and fallen walls, the straight walls and curving walls, the roads and tracks, the rough fell grasslands and neat valley-floor pastures, the Herdwick sheep, the valley-side farms and old cottages and even tracts of juniper make up a view with layers of natural and cultural interest from geological times to the present day and even glimpses of the future.

Pike Howe, at a height of 400 metres on the valley flank, is well positioned to allow the detail of this view to be seen. Not so high that the details of the scene are lost by virtue of distance and the grandeur of a vista; not so low that the landscape context of a feature is missed. Add to this the gentle thrill of exposure when standing on this rocky protrusion and you have a viewpoint of character that enhances the whole experience of the view. This is the author's fancy – a viewpoint that is an airy spot combined with a view with a story to tell. Guiding you to these places is what this book is about.

The airy spots from which 22 views are described in this book are named either "how" or "knott" on the Ordnance Survey map. Topographic features named "hows" and "knotts" are common in the Lake District landscape and are concentrated here more than anywhere else in the UK. In a sense they are themselves part of the Lake District's cultural landscape: both names have Norse origins and their shapes and position in the landscape help distinguish a Lakeland scene. They have an identity that is well recognised by walkers, scramblers and climbers and have been adopted more widely in the consciousness of the resident community in the names of scores of farms, hotels, houses and streets throughout Cumbria.

The 22 Lakeland views chosen for this book are all a delight to the eye, but they have not been chosen solely for this purpose. They have been chosen for the interesting historical journeys they display in themselves and also for what they tell us about ourselves and the way we see the landscape. For several thousand years humans have seen the Lake District landscape as a resource for living, then in the 18th century wealthy, educated visitors promoted an aesthetic way of seeing the landscape – the Picturesque – which led on to seeing the landscape as something worthy of conservation. In the 19th and 20th centuries the Lake District attracted more and more visitors from industrial cities and towns who saw the landscape as somewhere to find refreshment and enjoy both gentle and active pursuits. Today, in the 21st century, how we see the landscape has moved on again, with an ambition to ensure that a visually stunning Lake District is also an environmentally healthy one. The often–hard–to–see issues of climate change and ecological fragility are becoming recognised for the challenges they present. These historical shifts in the way the landscape is seen have not replaced each other but have become overlapping layers of landscape appreciation that are all present today. The views described in later pages provide illustrations of this.

This book is not an academic tome on landscape interpretation. It is an introductory guide to the stories behind a variety of Lakeland views that will, hopefully, prompt further exploration on the fells and in the literature. Whether you are reading this book with your boots on, sat on a rock, or with your slippers on, sat in an armchair, you will be taken to places familiar and off-beat. Hopefully your enjoyment of these places will be enhanced at least a little.

*Looking north-west from
Birk Knott towards Bowfell*

Making the Lake District Landscape

Before setting off on the descriptions of the 22 views in this guide, it may be helpful to many readers to start with the bigger picture of how the work of nature and the activities of humans have together created the present-day Lake District landscapes. After this broad overview, the view descriptions will fill in some of the detail and guide readers to see the views as examples that are typically "Lake District" but also as places where local geography and local human initiative have added a particular distinctiveness.

The Physical Landscape

It is sometimes claimed that the Lake District is a largely "man-made" landscape. This must come as a bit of a surprise to visitors from the cities: they have come to enjoy natural beauty and from their perspective the Lake District is nature personified. In no way is it man-made like a city in their eyes. Indeed, this man-made assertion is an exaggeration, as if the heaving masses of the fells and the watery expanses of the big lakes are incidental to the scene. Human activities may have put the make-up on the face of Lakeland but nature has provided the bone structure and the profiles that are the foundations for its arresting features and beauty. It is those *shapes* that are nature's primary scenic legacy – the shapes of the fells within which nestle the Lake District's distinctive mosaic of lakes, crags, pastures and buildings. The rock structures, elevations, slope profiles and spaces between these sculpted forms, from highest summits to deepest lake-holding basins, have determined soils, drainage, vegetation, wildlife, the opportunities for human settlement and ultimately the natural beauty that has led to the popularity and conservation of the Lake District.

The shapes we see in the Lakeland fells and valleys are not just a random, if pleasing, jumble; they are as they are for a reason. Understanding nature's scenic legacy through these shapes provides the base layer and entry point for understanding the later historic layers of human activity and impact on the landscape. Over the following pages, eight photographs illustrate what can be thought of as the "big shapes" in Lake District landscape. Understanding them does involve a foray into the worlds of geology and geomorphology (the study of landforms), but in an uncomplicated way that will, hopefully, not scare off too many readers.

Here is a view of the Derwent Fells looking west from Hause Gate, the col south of Catbells, south-west of Keswick.

What you can see here are relatively high fells, around 700 to 800 metres, soft in outline and displaying rounded summits leading out to spurs and narrow horizontal ridges. The valley slopes are relatively steep and generally smooth in profile with almost complete vegetation cover and prominent rock buttresses a rarity. Overall, there is a certain elegance to the shapes.

These fells are made of sedimentary rocks, known to geologists as the Skiddaw Group, that were laid down about 490 million years ago, making them the oldest in the Lake District. They were formed on the floor of an ancient ocean from the deposition of layers and layers of rock particles washed down from an eroding earlier landscape. The deposited material was subsequently compressed during later intense earth movements, covered by thick layers of different rocks and in places intruded by molten magma rising from deep below. Over time the overlying rocks were eroded away, presenting these older Skiddaw Group rocks at the surface for erosion, most recently during the Ice Age (see the glossary for a definition of the "Ice Age"). The earlier intense folding and proximity to the hot igneous intrusions hardened the Skiddaw Group rocks and made them relatively resistant to erosion, giving us today fells of some height and boldness in appearance. The distinctive smooth fell profiles stem from the relative homogeneity of the different rock strata in terms of their susceptibility to erosion, which makes sharp breaks of slope a comparative rarity. During the Ice Age, glaciers deepened the valleys and steepened their flanks but left few valley-side crags to mark their passage. The thinly bedded and contorted character of these sedimentary rocks means that, when they erode, they break down into small, thin and flat rock fragments (look at them under your boots when walking here) that readily move down slope, smothering pre-existing features and accumulating as concave slopes at the foot of the valley sides.

The next photograph is a view from Threlkeld Knotts at the north end of the Helvellyn range looking south-west across two tracts of land: High Rigg on the western side of St John's in the Vale and, beyond it, Castlerigg Fell on the west side of the Naddle valley.

Here you are looking at relatively low, broad and irregular terrain rising abruptly above the valleys and displaying a number of bold rock features, in particular the tilted benches on the flanks of High Rigg. The shapes here indicate origins very different from the Derwent Fells. Here we see rocks formed during a short and violent volcanic period of around five million years that started 450 million years ago as two continental plates in the earth's crust began colliding, similar to what is happening along the Pacific coasts of the Americas today. This initiated volcanic outpourings from a number of vents that completely covered the Skiddaw

Group rocks. In the first phase of this period, magma came to the surface in the form of extensive flows of molten lava. Then in a second phase a more viscous magma rose up, clogging up vents that then periodically exploded open in very violent eruptions depositing pyroclastic (literally "fire broken") material ranging in size from fine ash to large rock fragments. Rocks of considerable variety and complexity were formed, named by geologists the Borrowdale Volcanic Group.

This photograph of High Rigg and Castlerigg Fell exemplifies the first phase of volcanic activity, the outpouring of lava. The lavas often have widely spaced joints and fractures that make them strong and resistant to erosion, giving rise to poorly drained plateau surfaces and prominent rock faces as can be seen on Castlerigg Fell. The bench appearance on High Rigg arises from lava flows stacked one above the other, but separated by thin beds of pyroclastic and sedimentary material that have all been tilted and exposed to erosion. The different strata have resisted the erosive forces to different degrees, resulting in a very distinctive appearance that geomorphologists call "trap" (German for "step") topography.

Next is a view into the heart of the Scafell range from Crinkle Crags looking north-west. Here the shapes are very different again.

This is a high mountain area over 900 metres in height with rough, disjointed shapes, prominent crags, boulder-covered slopes and steep-sided valleys providing some of the most dramatic scenery in the Lake District. We are a long way from the smooth shapes of the Derwent Fells. Here we are looking at scenery made up of Borrowdale Volcanic Group rocks formed during the second and much more violent phase of volcanic activity. Vast quantities of material were erupted out of a volcano into the air and deposited over the land, layer on layer several kilometres thick, ultimately forming various kinds of pyroclastic rocks (such as consolidated fine ash deposits called "tuffs") and "volcaniclastic sandstones" (unconsolidated volcanic material eroded and transported into water to become sediments). After the bulk of the underground magma had erupted, the domed land collapsed back down in a chaotic way along a series of faults to create what geologists call a "caldera" – in this case, the Scafell Caldera. Later in geological time, around 400 million years ago, a massive mountain-building period, known as the Caledonian Orogeny, arrived during which all these volcanic rocks, together with Skiddaw Group rocks below and later sedimentary rocks above, were uplifted and folded to create an important Lake District structural legacy of folds with a south-west to north-east orientation. To gain some appreciation of the scale of these events, it is worth noting that the summit rocks of Scafell Pike, England's highest mountain (centre of the photograph), were once near the *bottom* of one of these folds.

The very rough and varied appearance of the scenery shown in the photograph can be explained in general terms by the heterogeneous character of the Borrowdale Volcanic Group rocks and their variable susceptibility to erosion, particularly during the Ice Age. More difficult to explain are the shattered rock features, upstanding "tors" and troublesome (for walkers) boulder fields of the Scafell and Bowfell summit areas. This is not typical of other summits in the Lake District. Ice and frost appear to be the main rock-splitting agents at work, but the issue amongst the experts is whether this happened when the peaks protruded above the ice covering the central fells or when they lay underneath it. This question returns when looking at the next photograph.

Next, overleaf is a view of the summit plateau and west-facing slopes of the Helvellyn range from near Stybarrow Dodd, looking north with Skiddaw and Blencathra in the background.

The shapes displayed here on the Helvellyn range are very different from those of the Scafell area, but surprisingly these fells are also formed from rocks of the Borrowdale Volcanic Group. Here we see volcanic rocks displayed, not as spectacular crags, but as smooth undulating plateau surfaces below which are smooth convex slopes curving down steeply into the glacial trough of the Thirlmere valley. If geology is not the explanation for the differences between the two ranges, the character of their respective glaciations probably is. The gentle profiles of the summit plateau might suggest smoothing by moving ice, but in fact evidence to indicate the passage of ice needs to come in the form of roughened ground, striations in rock surfaces and glacial erratics, none of which has been found here. This might seem strange when the whole Lake District is described as a glaciated landscape.

During the Ice Age the ice conditions were not constant and understanding how different conditions could have led to particular landscapes and landforms is a complicated business. During the coldest periods the ice was frozen to the ground and little erosion took place under it; in less cold conditions the ice, assisted by meltwater, moved and erosion took place. At some times the highest summits protruded above the ice, while at others the whole district was completely covered. So perhaps the best way to view it is that the landscapes we see today are the product of average ice conditions during the glacial period. But even then, so-called average ice conditions probably differed between the west of the Lake District, where the prevailing winds brought plentiful snow, and the more sheltered east. This is thought to be the key to understanding the difference in appearance between the Scafell and Helvellyn ranges. With less precipitation in the eastern Lake District, the Helvellyn range (and the High Street range) is thought to have had a relatively light-touch glaciation over the summits and less vigorous valley deepening by glaciers than in the west. This suggests that the appearance of the Helvellyn range in the photograph might be similar to the Lake District landscape prior to the onset of the Ice Age. If this is the case, it makes the point that without the Ice Age the Lake District would be a much blander landscape than the one we enjoy today.

The next photograph is a view of Eskdale from near Eskdale Green looking north.

Here the valley flanks display prominent bulbous rock outcrops that give a distinctive character to the scene and above them lie wild, open moors interspersed with smoothed rock outcrops and wet hollows. This is granite country shaped by south-moving ice. At the end of the short volcanic period around 450 million years ago, large volumes of molten magma intruded into the Borrowdale Volcanic Group rocks, metamorphosed the rocks they came into contact with and then slowly cooled underground to create coarse grained granites. The most extensive granites visible on the surface today are the Eskdale and Ennerdale intrusions, which extend from a huge chamber of granitic rock that is thought to underlie the whole Lake District and beyond. Granite can usually be identified by non-geologists because of its large crystals (the result of the slow cooling underground) that in this part of the world give off a light pink hue. If you can see this when passing walls and buildings you will probably be able to get your eye in to discern the rock in the protruding bare valley crags that stand out from the surrounding vegetation. The granite has well-developed joint systems that when weathered and eroded give the crags a distinctive blocky appearance and provide the alignment for numerous gullies and becks.

Next is a view from Latterbarrow looking north-west over some of the fells of High Furness with the volcanic Coniston Fells in the background.

This is very subdued terrain by Lake District standards, barely reaching 300 metres in height, with long, low ridges and shallow valleys substantially covered in trees, making it difficult to appreciate the character of the land. This is country based on a series of sedimentary rocks, known to geologists as the Windermere Supergroup, that stretch across much of south Cumbria. They were laid down during the Silurian period some 420 million years ago on top of the volcanic rocks that by this time had been subject to folding and erosion. The Lake District had subsided into an ocean basin after the volcanic period and the changing depths of water and varying sources of sediment led to a wide variety of sedimentary layers of limestone, muds, silts and sands being laid down. Like the older rocks below them, these rocks were folded into north-east to south-west structures during the Caledonian Orogeny. This in due course determined the orientation of the low rocky hills and scarps of more resistant rocks and the low depressions of softer rocks. The rocky hills and scarps add some welcome interest to an otherwise undemonstrative landscape.

The six photographs looked at so far have shown differences in the shapes of the fells and served to introduce the main rock types of the Lake District. Here is a geological map that shows the full extent of the different rock types that have been discussed. The map also shows younger rocks, mainly limestones and sandstones, which were laid down on top of the rocks of the Windermere Supergroup. During periods of uplift some 300 million years ago (known as the Hercynian Orogeny) all these rocks were raised into a dome. As the top of the dome was gradually eroded, the uplifted older rocks were exposed in the centre, surrounded by a ring of younger rocks; a layout that can be seen on the map.

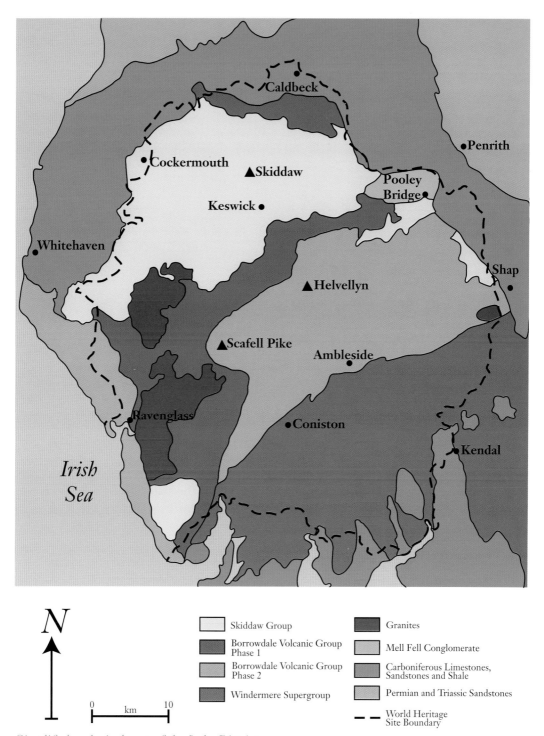

N

Caldbeck

Penrith

Cockermouth

▲Skiddaw

Pooley
Bridge

Keswick ●

Shap

Whitehaven

▲Helvellyn

▲Scafell Pike

Ambleside

Ravenglass

● Coniston

Kendal

Irish
Sea

	Skiddaw Group		Granites
	Borrowdale Volcanic Group Phase 1		Mell Fell Conglomerate
	Borrowdale Volcanic Group Phase 2		Carboniferous Limestones, Sandstones and Shale
	Windermere Supergroup		Permian and Triassic Sandstones
		- - -	World Heritage Site Boundary

0   km   10

Simplified geological map of the Lake District

The upstanding fells are obvious "big shapes" in Lake District landscapes, but equally so are two other features: the deep, glaciated Lakeland valleys and the huge fellside hollows known as "corries". The next photograph is taken from Warnscale Bottom looking north-west down the Buttermere valley.

Here we see the classic shape of a Lake District glaciated valley with steep sides and a flat valley floor occupied by a lake. Big lake-filled valleys like this are the essence of Lake District landscapes. It can be seen from any map that the main valleys and their lakes appear to take a radial form. However they do not, as is sometimes suggested, radiate from a central point but rather flow away from a west-to-east watershed that winds its way from west of Pillar to east of High Street. Exactly how the pattern of river drainage came about remains an area of discussion amongst the experts. The curiosity is that the river systems do not follow the broad south-west to north-east structural trend of the underlying geology established during the Caledonian Orogeny and a number of theories have been put forward to explain this. Current best thinking is that weaknesses in rock structures associated with faults and fractures created during the mountain-building episodes were the primary determinants of river patterns, albeit with some modification during glaciations.

During the Ice Age, ice flowed outwards from the high ground of the central Lake District usually along the lines of the pre-existing river valleys that then became modified in shape. During the early and late stages of the glacial period, the valleys would probably have looked similar to today's Alpine valleys with a slow-moving glacier filling the valley bottom and the valley head and interfluves partly free of ice. During the coldest periods when an ice sheet covered the whole District it is thought that streams of ice, moving significantly faster that the ice mass as a whole, would have occupied the valleys and been particularly effective agents in deepening and widening them. As the Ice Age came to an end the profiles of the valleys we see today gradually emerged: crags unsupported by ice collapsed and in the cold periglacial conditions, frost-shattered rocks fell down the valley sides. Vast amounts of debris accumulated at the foot of slopes, masking any semblance of the textbook 'U'shaped valley cross-section (the term "glacial trough" is probably more appropriate). At the same time the elongated rock basins in valley floors, over-deepened by the tremendous erosive power of the ice, filled with water to create elongated lakes, commonly referred to as "ribbon lakes".

This final photograph is a view of the eastern side of the Helvellyn range immediately south of Helvellyn summit and Striding Edge.

Here we are looking into a series of spectacular rocky hollows commonly called "corries" (but termed "cirques" by geomorphologists and named "coves" or "combes" in the Lake District). These large armchair-like shapes are dramatic features in the landscape and can

be seen across Lakeland, particularly in the hard resistant rocks of the Borrowdale Volcanic Group. They were gouged out by glaciers that developed through the accumulation of snow blown off the summits by south-west winds into pre-existing hollows that faced between east and north, shadowed from the heat of the sun. The ice eroded and enlarged the hollow, assisted by a rotational slipping movement within the confines of the bowl. In some cases the ice within the corrie would have been augmented by ice spilling over from the headwall above and expanded and flowed down-slope to join a valley glacier. It is unlikely that corries of the size shown in the photograph were created solely during the most recent Devensian glacial period, particularly bearing in mind that the Helvellyn range is thought to have had a light-touch glaciation during this period. They are probably the accumulated product of repeated glacial phases extending over the Quaternary period (the last 2.6 million years), the most recent being a short cold period known as the Younger Dryas which began 13,000 years ago and lasted for about 1,200 years. During this latter period, localised small glaciers (but not valley glaciers) became re-established, further hollowing out corries and leaving some of the more prominent glacial moraines seen across Lakeland.

Finally, to conclude this section, here is a brief sketch of the transition from nature's shaping of the landscape to the creation of an environment suitable for human habitation. As the glaciers gradually retreated at the end of the Ice Age the huge quantities of eroded rock debris they were transporting within them and on their surfaces were dumped as extensive layers of glacial till over wide areas. These were shaped into moraines at points where glaciers temporarily paused or carried away to be deposited widely by meltwater as glaciofluvial deposits. Immature braided streams spread material across the valley floors, sometimes filling in lakes only recently formed in ice-gouged basins to create the flat valley floors that are so noticeable in many Lake District valleys.

The deposited glacial and fluvial gravels, sands and silts in due course created skeletal soils of purely mineral material. The introduction of organic matter to develop soils came with the growth of plants starting with colonisation by lichens and mosses followed by grasses, arctic plants and scrubs, like juniper and dwarf willow. Then trees gradually established themselves firstly with birch, hazel, pine and then oak and elm. With the accumulation of forest humus, a relatively fertile base-rich forest soil developed up to about 650 metres. It was into this new post-Ice Age environment that the first intrepid humans ventured.

The Human Landscape

The story of the human role in shaping the Lake District's landscape is about how the legacy of the physical landscape and its natural resources were utilised to create the particular configuration of woodland, pastures, farmsteads, walls, settlements and roads we see in every view today. It is the history of evolving practices of land use and management in response to changing community needs and ambitions. Land management was initially simply about providing for the daily needs of food and shelter and then later to provide agricultural and other products for trade. More recently land management has also sought to protect and enhance the landscape's aesthetic beauties and now, in the 21st century, to help secure the landscape's long-term environmental health. The objectives of land management in the Lake District today have become multi-layered and complex.

The following pages give a brief historical overview of past and present human land-management practices within the Lake District as a whole with the intention of providing a context for the more detailed look at the 22 views that follow.

Early History and Clearing the Forests

The first humans to enter Cumbria in significantly sized groups arrived during the Mesolithic period from about 8000 BC onwards when evidence suggests that groups of hunter gathers began to occupy the south-western coastal areas, such as the Eskmeals area, south of Ravenglass. These peoples exploited the resources of the sea and estuaries, hunted wild animals and gathered food products from the expanding forests. However, they were living in a post-glacial physical environment that was still evolving. Some coastal areas were overwhelmed as sea levels rose with the continuing melting of the ice sheets, displacing early humans and drowning newly established forests; only for the sea to later retreat as the land, released of the weight of ice, rose and provided fresh opportunities for human occupation. By around 5000 BC the present river systems were largely established, soils well developed and forest cover more or less complete, with oak and other deciduous trees on lower areas and birch and pine on higher ground up to around 750 metres. Above was open grassland. This image of the Lake District largely under forest at this date marks the starting line for the human creation of the cultural landscape.

Remnant of ancient oak wood surrounded by heather in full bloom, Keskadale, Newlands, from Aikin Knott

The gradual destruction of the original forest cover has been by far the biggest impact of human activity on the Lake District's landscape and ecology. Where there are high levels of rainfall, as in the Lake District, soils are vulnerable to leaching and fertility levels can only be retained with a vegetation cover that constantly renews the base supply in the soil surface layers. A combination of diminished forest cover, which no longer sufficiently recycles soil nutrients, and a heavy rainfall regime leads to soils becoming acidic and lacking fertility; something that in due course came to pass.

From about 4000 BC there is evidence that areas of forest were being destroyed by fire. Much of the information we have about the prehistory of the Lake District comes from analysis of pollen samples taken from bogs and the edges of tarns. These provide information about changes in vegetation cover that in turn reflect the effect of natural events and early human activities. The pollen record provides us with evidence of charcoal around 4000 BC, suggesting the burning of forest from natural events or possibly by Mesolithic peoples to create clearings to attract wild animals. A stronger indication that human activities caused forest loss is the evidence in the pollen record of a marked decline of elm and increase in grasses after 3000 BC. This is thought to mark the impact of recently arrived Neolithic farmers in the area and the start of the loss of the forested landscape. The Neolithic peoples hunted, fished and kept cattle, sheep, goats and pigs and made clearings in the forests for growing crops and grazing areas. This is the period when the famous stone axes made from the flinty, volcanic rock found in the high fells of Langdale were first put to use. The population was still fairly mobile but when clearings were abandoned they would not naturally regenerate if grazing by animals continued. This is also the time when stone circles started to appear within the cleared, open landscapes, such as the famous example at Castlerigg, near Keswick.

By the end of the Neolithic and into the Bronze Age, around 2000 BC, the population was becoming more settled and, with a warmer climate, expanding onto lower fells up to around 300 metres in places like Moor Divock near Pooley Bridge, on Eskdale Moor and on Stockdale Moor and Town Bank, south of Ennerdale Water. In such locations the land was cleared of trees and stones and put to agricultural uses, while the river valleys still remained wet and forested. Evidence of occupation by Bronze Age peoples at such locations comes in the form of small hut circles, field walls, clearance cairns, burial cairns, small stone circles and standing stones. By the late Bronze Age, around 1000 BC, it appears the climate deteriorated and many of the higher settlements were abandoned. By now the cleared areas were showing signs of increasing soil acidity and erosion with the old forest soils developing into peat, particularly on the flatter topped ridges.

During the Iron Age, from 800 BC onwards, agricultural activity shows evidence of organised land and woodland management with settlement on the lower fells and increasingly in the valleys. The evidence seen in the landscape today includes the remains of settlements, such as at Glencoyne beside Ullswater, and possibly some small hillforts (although some described as Iron Age may actually date to the immediate post-Roman period).

Remains of a hut circle at Tongue How within the largely Bronze Age landscape of Stockdale Moor and Town Bank

The Romans arrived in the Lake District in about 100 AD and overall made little impact on the landscape. Cumbria was a fringe region to be managed, rather than occupied and exploited. The main Roman legacy is the remains of a series of forts and a road system that connected them, together with civilian settlements at Ambleside and Ravenglass. The growing of crops, such as cereals and hay, and the management of cattle and sheep provided the livelihood of the Romano-British population. During the latter part of the Roman occupation upland deforestation intensified with the expansion of cereal production and the demand for wood for building. However, there was little clearance of the woods in the central valleys and the swampy valley floors remained undrained. In the centuries following the Romans' departure at the beginning of the 5th century, the little evidence there is suggests a mixed picture with woodland clearance, more intensive agricultural activity and settlement up to 300 metres in some places and in others, land abandonment associated with deteriorating soil conditions arising from over-cultivation and over-grazing. These were probably somewhat unruly times and some hillforts, such as Castle Crag, Mardale and Castle How, Bassenthwaite, probably date from this period.

In the 10th century, colonisation by Norse from Ireland got underway in what appears to have been a relatively peaceful process with the new settlers seeking out more sparsely populated areas, particularly in the valleys. They extended woodland clearance for sheep and cattle pastures, introduced the ancestors of the Lake District's traditional Herdwick sheep and had a lasting impact on settlement and farming patterns. They developed transhumance (the movement of livestock to summer pastures) as a feature of farming practice, with temporary summer camps (known as "shielings") on higher land and at valley heads that in some locations later became permanent settlements. It was during this period leading up to the arrival of the Normans that the valleys were substantially cleared of forest and settled, accelerating the process of destruction of much of the extensive native oak woodland.

Landscapes of Farmsteads and Enclosure

Norman influence arrived in Cumbria in the late 11th century when a series of baronies were established and feudalism brought in new land-management regimes to the occupied lands of the Lake District. The Norman baronial estates, based around the periphery, extended over much of the Lake District with huge areas of private "forest", a term still appearing on modern Ordnance Survey maps, as in the case of Skiddaw Forest. This term did not mean an area was fully forested, but rather that it was reserved for the preservation of game animals for hunting. Over time the feudal lords gave priority to stock grazing rather than hunting and vaccaries (cattle/dairy farms) were established at some valley heads. For a while the lords retained occupation of their lands at the valley heads but by the end of the 13th century much of their land in the valleys was divided and leased out to tenant farmers who benefitted from feudal rights within the forest in return for service, notably military service to defend the border areas from raiding Scots.

However, not all land in the Lake District remained under the control of the feudal lords during the medieval period. The abbeys at Furness, Calder and Shap and the priories at Cartmel and Conishead, as well as Fountains Abbey in Yorkshire, were founded in the 12th century and granted very extensive lands in the Lake District by the feudal lords. These religious houses managed their lands and their tenants in accordance with their respective traditions, took over or established numerous vaccaries, such as at Brotherilkeld in Eskdale, and built granges as focal farmsteads in their territories, such as at Grange in Borrowdale. By the time of their dissolution by Henry VIII in the 16th century, they had been substantially responsible for turning the open fells into sheep pastures and establishing commercial sheep farming as a mainstay of the Lake District economy.

It was during the medieval period that the Lake District's distinctive settlement pattern of individual and clustered farmsteads was established. At the valley heads the early shielings and former vaccaries became farmsteads. Within the narrow valleys, where good flat land was in short supply, farmsteads were strung out just above the valley floor on south-facing fellsides adjacent to spring water supplies. Around the mountain periphery, where the local topography allowed larger areas of well-drained land, groupings of farmsteads were established with communal cultivated open fields adjacent.

Within a typical Lake District upland valley the landscape would have consisted of a common open field of cultivated strips and meadow land on the valley bottom (known as the "inbye" land) surrounded by a continuous wall (the "ring garth") that protected it from damage by wandering stock grazing on the open fells. The open fells were common land where specific rights to pasture animals, cut peat for fuel, cut bracken for thatch and animal bedding, and collect certain categories of wood for fuel and fodder were held by tenants of individual farms. On the fells stone shelters ("bields"), sheep pens, pinfolds and peat huts appeared to support the shared use of this land. From 1450, after a period of plague and economic depression, a growing population needed additional land for cultivation and stock rearing and land immediately upslope of the ring garth was enclosed as walled "intakes" by individual farmers and farmers working together. These valley enclosures (that is, the ring garth and intakes) were the first stage of building the prominent walled landscapes we see today.

Although the population was growing there were no substantial settlements in the Lake District itself during the medieval period. The main market towns were outside the District at Kendal, Penrith, Ulverston and Cockermouth and within the District only Hawkshead and Keswick served as significant market centres. A number of small settlements like Pooley Bridge, Ravenglass, Hesket Newmarket and Staveley were granted market charters, but this did not stimulate significant growth in population or function.

By the late 16th century the monasteries had been dissolved and their lands sold off, and with most of the landowning gentry living outside the core Lake District a very independent-minded breed of tenant farmers rose in prominence and influence. Their lands were held with customary rights that provided security of tenure for the tenant and his descendants linked to the obligation to provide military service. Following a legal challenge this right was confirmed in 1625, which then gave these yeoman farmers the security to make substantial investments in their farms. It was in the late 17th and early 18th centuries that many of the farms in the Lake District were rebuilt in stone. These yeomen farmers increased their wealth and became a distinctive Lakeland farming society, later to be romantically termed "statesmen" (from "estatesmen") by visiting travellers and the first tourists. The influence that these traditional farmers exercised over the evolution of the Lake District's agro-pastoral farming landscape did much to create the distinctive qualities of the landscapes we see today.

Nevertheless, the period of the yeoman farmers was a period of change in land management partly through private agreements between individual yeoman farmers and partly through

Townend, Troutbeck; a statesman's house dating from 1626, owned by the National Trust

the decisions of the manor courts. The common open field system gradually disappeared and crop cultivation on the valley bottoms was within individually owned small fields separated by newly built walls or hedges. On the lower fells new walls were built by owners of the more prized cattle pastures, while on the high fells individual farms obtained exclusive grazing rights to defined areas of common and their sheep flocks learnt to stay in and graze a particular area of fell, becoming what is known as "hefted". This was the second stage in the evolution of walled enclosures in the typical upland Lakeland valley.

Eventually, by the second half of the 18th century, triggered by large-scale science-based improvements in agriculture and rural industries nationally, major changes in the rural economy brought the era of the statesmen farmers to an end. Wealthy individuals with agricultural and financial interests inside and outside the Lake District bought up and amalgamated farms, while many others were simply abandoned. The General Enclosure Acts from 1801 onwards and the decisions of the enclosure commissioners brought further impetus and order to rural change with new measures to improve existing cultivated land, prevent over-grazing of upland commons, prevent soil erosion and excessive removal of peat, and bring poorer land into production.

Although the Acts led to the enclosure of some open-field arable land with hedges and walls, the main landscape impacts were the enclosure of large tracts of common land on the fells. Here commoners' rights were removed, ownership broken up and land "allotted" to new owners whose names sometimes appear on today's Ordnance Survey maps. Planned landscapes were created with large walled fields, new access roads and improved land drainage, as well as new farmsteads that introduced further changes to the settlement pattern. The enclosure walls introduced long, straight lines and rectangles into the landscape and were typically constructed from field and quarried stone, although in parts of the District local iron industries provided iron fencing. This was the third and final stage of building the walled enclosures so prominent in the Lakeland landscape.

These new landscapes of Parliamentary Enclosure are most readily recognised outside the upland core of the Lake District where there were significant changes to the pattern of settlement, buildings and roads. In the central Lake District, examples of these rectilinear landscapes can be seen, but the constraints of the physical environment and the legacy of statesmen farming traditions impeded the scale of change and, in particular, the scale of loss of common land. With the coming of the Commons Registration Act 1876 the pressure to enclose was replaced by measures to improve the management of common land in the uplands so that today Cumbria retains more common land than any other part of England.

Stooks of corn, Hollows Farm, Borrowdale; early/mid 20th century

Following these dramatic changes in the Lake District's agro-pastoral scene, further landscape change continued during the 20th century with the widespread abandonment of crop cultivation in the valleys as improved transport allowed foodstuffs produced in more favourable areas to be imported. Distinctive bright-green sheep pastures came to typify the appearance of the valley floors set amongst the walls of an earlier era. Within such valley landscapes now lie the history, traditions and legacies of over 1,000 years of distinctive agro-pastoral farming. Most of the walls, farms and associated buildings stand on the foundations of their ancient predecessors and the farming families themselves are supported by the handed-down knowledge and traditions of their forebears. Over centuries the ewes and lambs of Herdwick flocks have been moved up and down between the most nutritious pasture of the inbye land, the intakes and the poorer fell land in rhythm with the seasons and the animals' life cycle. Farmers have exercised their ancient rights to graze their flocks on the common unenclosed fells where the sheep have learnt their particular heaf. What has evolved is a unique communal farming system dependant on everyone playing their part, creating strong social traditions, perhaps most publicly expressed through the annual valley sports and shows. It is these stunning pastoral landscapes that inspired the idea of seeking World Heritage status for the Lake District.

The Industrial Landscape

It comes as a surprise to many visitors that the Lake District has an important industrial history and landscape. The key raw materials were rocks, minerals, wool and wood that were processed by the power provided by charcoal and fast-flowing water. The peak came during the 18th and 19th centuries when the growing industrial cities and towns, accessible with improved transport, generated a huge demand for various products derived from these raw materials.

The working of the Lake District's rocks goes right back to Neolithic times with the working of volcanic rocks to fashion valuable stone axes that were widely traded in other regions. A major "stone-axe factory" was discovered high up on the side of Pike O'Stickle in 1947 and subsequently at other locations in the central fells. However, it was not until the 17th century that the working of the Lake District's rock resources really began to get underway with the widespread rebuilding of farmsteads and other buildings in stone – sometimes referred to as "the Great Rebuilding". Quarries for building stone and slate were opened up just about everywhere and today the appearance of multitudes of buildings across the District reflects the source geology. Building programmes in the expanding cities and towns in the 19th century led to a huge demand for roofing slates, with production from the area's many slate quarries peaking in the 1890s. A small number of slate quarries still operate, such as at Elterwater and Coniston.

Mineral-vein deposits containing lead, copper, zinc, silver, iron and several other minerals are found in many parts of the District. It is possible the Romans mined copper at Coniston and during the medieval period iron ore was mined, with names such as Ore Gap, Red Tarn and Red Gill giving a clue to the whereabouts of the red haematite deposits. The exploitation of the Lake District's mineral resources accelerated in the late 16th century during the reign of Elizabeth I when copper and lead mining and smelting developed in the Keswick and Caldbeck areas. Although the smelting of copper and lead eventually moved to locations outside the District, the smelting of iron continued into the early 19th century using iron ore from Low Furness and vast quantities of local charcoal for fuel. The demands for all metals increased with national industrialisation and urban expansion in the 18th and 19th centuries and water power was harnessed in a big way in many mining operations to lift ore and work crushers and pumps; and with respect to iron working, to power bellows and hammers for furnaces and forges.

The Lake District wool industry grew during the medieval period with the monasteries at the forefront of its development. Kendal became a major wool town supporting the growth of cloth manufacture in the District, initially on cottage-industry lines. The town was granted its borough charter in 1575 and controlled all wool-finishing trades and marketing. Water-driven wool fulling mills had been part of the landscape from medieval times, but by the late 18th century the coming of the factory system and technological developments had led to an increase in the scale of production; wool processing and cloth manufacture moved away from the farms and transferred to new water-powered woollen mills. This was one example of a source of income being taken away from the families of the yeoman farmers, helping to undermine their livelihoods.

The Lake District's natural woodlands had been a vital resource from earliest times, but it was the managed "coppice" woodlands, particularly in the southern Lake District, that became the resource for large-scale charcoal making and a host of woodland industries from the medieval period onwards. Coppicing involved cutting a young tree down to just above ground level so that new shoots would grow up from the stumps to form timber poles some six metres in height. These were harvested to provide wood that could be readily handled, worked and transported. Coppice wood was used to make charcoal and provide raw materials for many industries like brewing, textiles and farming as well as for buildings and furniture. Making charcoal had a particularly dramatic effect on woodland loss until woodland management regimes, including new coppice planting, were introduced. Charcoal was particularly in demand to smelt iron, which in medieval times was done in simple unpowered smelting hearths known as "bloomeries". By the 18th century, water-driven ironworks using early blast furnace technology had arrived in the Lake District and with them the demand for vast quantities of charcoal. The other woodland industries also harnessed water power, for example in the manufacture of vast quantities of wooden reels and bobbins in the early 19th century to meet the demands of the Lancashire cotton industry. This led to the opening of over 60 water-powered bobbin mills mainly in the High Furness area.

Charcoal-fuelled iron furnace at Duddon Bridge

In the 20th century all these industries declined as steam power succeeded water power, production transferred to the industrial cities and alternative sources of raw materials from around the world were exploited. Today it is spoil heaps, overgrown quarries, old mine entrances, derelict stone buildings and a handful of operational slate quarries that are the most obvious reminders of this industrial history, but there are also many less obvious reminders worth looking out for, such as old coppice woods and bloomery sites. These industries were not Lakeland-pretty in their heyday and neither are their relics, but they have been important contributors to the evolution of the present-day cultural landscape, not least because of the crucial supplementary income many of them provided to sustain Lakeland farming families until the arrival of the tourists.

New Ways of "Seeing" the Landscape

From prehistoric times until the 18th century the land in the Lake District had been seen in a hard practical way as a resource from which food and all the goods for everyday living and trading were obtained. This was the accepted way of seeing the landscape and its appearance in the 18th century reflected this purpose. It was a landscape of agricultural endeavour interspersed with small pockets of intense mining and industrial activity set within vast tracts of largely unvisited fells. But in the 18th century everything started to change, not on the initiative of the local inhabitants, but through the coming of a new way of "seeing" the land by educated, wealthy outsiders. In a most extraordinary way, direction over how the Lake District land should be used and managed was taken out of the hands of the people who owned and worked it. It was to be a revolution.

Early-18th-century travellers on exploratory tours of the country had seen little to excite them in the Lake District. The London-based novelist and journalist Daniel Defoe was typical when in 1726 he described the District as "eminent only for being the wildest most barren and frightful of any I have passed over in England", doing nothing to suggest a visit was worth the effort. During the second half of the century, however, perceptions started to change.

On their European Grand Tours, the educated wealthy classes had become accustomed to viewing and appreciating dramatic Alpine scenery. Their understanding of scenery, particularly mountain scenery, evolved, influenced by greater contact with it and also by wider contemporary intellectual discussion about the meaning of and responses to the aesthetic ideals of the "beautiful" and the "sublime" (the latter equating with "awesome" in today's language). These two aesthetic ideals were a reaction to the rationalist thinking of the Enlightenment and enabled non-rational, emotional responses to be brought to the appreciation of landscapes. The Lake District became a regular stop on their itineraries and their reports on the wild scenery used this terminology and freely expressed the emotions evoked (sometimes in hyperbolic terms in the case of some over-excitable writers).

The Lake District and other British landscapes came to be seen not only as working countryside but also in terms of their aesthetic qualities and the emotions they evoked. A movement, known as the "Picturesque", emerged and soon grabbed the interest of the educated classes. One of its most prominent promoters was William Gilpin who toured the Lakes in 1772. Gilpin defined the Picturesque as "that peculiar kind of beauty which is agreeable in a picture". He set out some rules about textures, composition, perspectives, and light and shade that were desirable within the framed view. The old classical principles of symmetry, perfect proportions and clarity were rejected in favour of accidental irregularity, intermingling of visual stimuli and rusticity. The Picturesque became the taste of the times.

Many guide books were published to introduce readers to the Picturesque delights of the Lake District. One of the most widely read in the early days was Thomas West's *A Guide to the Lakes* published in 1778. In his tour of the District he picked out 21 "stations" (viewing points) and instructed his readers on the visual qualities of each view. The aesthetic delights were everything and searchers for the best scenes began visiting in increasing numbers.

This new way of looking at and appreciating the landscape chimed with the European intellectual movement of Romanticism in the first half of the 19th century. The movement

Claife viewing station, built beside Windermere in the 1790s

was a reaction to industrialisation and a science-based view of the world and fostered new alternative ideas about human emotions and relationships with nature. The Lake District was the major source of inspiration for the Romantic movement in England, led by the poet William Wordsworth and fellow poets Samuel Taylor Coleridge and Robert Southey (known together as the Lakes Poets) and attracting many other notable literary and artistic figures. The Lake District and other landscapes were to be appreciated for their intrinsic qualities and the emotional responses they triggered, bringing individuals a sense of self and freedom. In many respects these pioneering views can be seen as the forerunner for what today we call environmentalism: living in harmony with nature. But not everyone was seeing the Lake District in quite the same way – in particular, the growing industrial cities of northern England.

Wealthy industrialists and business owners from the northern cities, alerted to the District's natural beauties, started a fashion to build substantial residences, "villas" as they came to be termed, with designed gardens in prime lakeside locations. Then, as the 19th century progressed, a wider spectrum of people from the industrial towns with some leisure time and money to spend also took up the idea of visiting. Pretty views, fun boat trips on the lakes and, for the more energetic, fell walking were the appeal. The railway companies spotted the opportunities and the first line into the District was opened in 1847 bringing people to the village of Birthwaite which later grew into the new town of Windermere. Seen through the eyes of those with an entrepreneurial spirit, whether living in the District or elsewhere, the Lake District clearly had business potential. Keswick, Ambleside and Bowness developed tourism economies and expanded in size with hotels, guest houses, new residential housing and various tourist facilities and attractions. The appearance and character of the Lake District was changing throughout the 19th century.

But it was not just private individuals who were seeing opportunities in the Lake District. Manchester Corporation desperately needed a new source of clean water to help relieve squalid and unhealthy living conditions in the city and saw an answer to its predicament in these northern valleys and lakes. In 1877, it sought and was eventually granted parliamentary powers to build a reservoir in the Thirlmere valley to provide a reliable and plentiful supply of water to the city.

The 19th century showed clearly that different people had different ways of seeing the Lake District landscape and different ideas about how this corner of England should be used. Furious public debates, at the national level as well as locally, were provoked by the proposals to bring the railway to Windermere and to build the Thirlmere reservoir. At stake was which of the different views should prevail. The one that eventually prevailed was that the Lake District landscape should be seen as a special place of great natural beauty and a vital source of human inspiration and refreshment, and that it should be highly valued for those qualities. Surprising

View of the Thirlmere valley before the reservoir was built

Thirlmere and Saddleback – near Keswick   Abraham's Series No. 201.   Keswick

though it may seem to us today, this was in its time a novel way of seeing the landscape and it led on to another equally radical idea – that of landscape conservation.

The Idea of Conservation

Concerns about what was happening to the Lake District landscape started emerging in the late 18th century. Small local preservation and campaigning groups came into existence to resist the closure of public footpaths and harmful schemes of various kinds but, like today, effective campaigning needed an articulate figurehead and a good slogan. The initial provider of this combination was the poet William Wordsworth. In 1822, Wordsworth published a pocket-sized tour guide entitled *A Guide through the District of the Lakes* that was rather more a manifesto about landscape management than a visitor's guide. Wordsworth's wish was that he would be joined by others to "testify that they deem the district a sort of national property, in which every man has a right and interest who has an eye to perceive and a heart to enjoy". Wordsworth, through his own campaigning, his poetry and other writings, established and gained acceptance for the idea of conservation of the Lake District and with those words "a sort of national property" he provided the rallying slogan. The idea of conservation became established in the national consciousness; however, it would require practical action by many individuals over many decades to turn the idea into reality. The conservation movement gradually developed along two complementary lines: campaigning and acquiring land for protection.

Wordsworth, his fellow Lake poets, the prominent art critic and social thinker John Ruskin and other notables campaigned individually and collectively during the 19th century against specific major proposals, notably further railway schemes, while local groups continued their local battles about access, tree felling and such like. After the battle over Thirlmere, the Lake District Defence Society was set up in 1883 and other District-wide preservation groups came and went over the years.

Alongside case-by-case campaigning, it was realised by Wordsworth and others in the early part of the 19th century that the surest way of protecting valued areas of Lake District landscape was for the right person or body to own them. A number of wealthy individuals, such as the enlightened Leeds industrialist John Marshall who acquired estates in several valleys, took it upon themselves to do just that and by the end of the 19th century even the notion of state ownership of the Lake District was raised in some quarters. What did emerge in 1895, however, was the creation of the National Trust – or the "National Trust for Places of Historic Interest or Natural Beauty" to give it its full title. The creation of the Trust provided the opportunity to acquire land by public subscription, accept gifts of farms and other land and enter into covenants with landowners who wished to safeguard their land in perpetuity. In the early decades of the 20th century, people like the prominent historian G.M. Trevelyan and the writer and farmer Beatrix Potter made very significant donations of farms to the Trust for safekeeping. The Trust now owns nearly a fifth of the land in the Lake District.

Yew Tree Farm, Coniston; bequeathed to the National Trust by Mrs B. Heelis (Beatrix Potter)

The efforts to turn the idea of conservation into practical reality continued into the 20th century and new challenges soon arrived. Wordsworth's idea of the Lake District being "a sort of national property" was not yet widely accepted; it was still largely a free-for-all. After the First World War, the Forestry Commission was set up in 1919 to boost the nation's timber supplies and provide rural employment. The planting of swathes of incongruous, regimented blocks of forest in Ennerdale and elsewhere and plans for a Hardknott Forest Park in the upper Duddon and Esk valleys caused outrage that ultimately led to an historic agreement with the Commission in 1936 that 300 square miles (777 square km) of the central fells should stay free of state forestry – a commitment that still stands today.

By the 1930s, the conservationists' twin strategies of campaigning and acquiring land were having some effect and calls for Parliament to agree to the creation of statutory national parks in the Lake District and elsewhere were gaining momentum. The mass trespass on Kinder Scout in the Peak District in 1932 was an important moment in moving hearts and minds but it was not until 1949 that opposition was finally overcome and legislation passed for the setting up of national parks in England and Wales. The Lake District finally became a national park in 1951. The National Park Authority (the Lake District Planning Board as it was named initially) was required to conserve the Park's natural environment and encourage the public's enjoyment of it and, most importantly, took over planning powers from the local authorities. There was now in place a single body with a statutory Lake District-wide conservation remit.

However, the efforts to frame the way the Lake District should be seen and valued did not rest there and, towards the end of the 20th century, the goal to have the area recognised on the international stage as a World Heritage Site was pursued. This finally became a reality in 2017 when UNESCO inscribed the Lake District onto the World Heritage list as a cultural landscape and adopted a Statement of the District's Outstanding Universal Value. In summary, three inter-linked aspects of the Lake District are recognised that give it global value:

> **Identity:** A landscape of exceptional beauty, shaped by persistent and distinctive agro-pastoral traditions which give it special character

> **Inspiration:** A landscape which has inspired artistic and literary movements and generated ideas about landscapes that have had global influence and left their physical mark

> **Conservation:** A landscape which has been the catalyst for key developments in the national and international protection of landscapes.

In 2015 the shepherd and life-time Lake District resident James Rebanks wrote in his book *The Shepherd's Life* "There are places where it doesn't feel like it's ours any more, as if the guests have taken over the guesthouse". This is precisely what has happened; it is not only that the "guests" have moved in to buy up houses for second homes and holiday lets; they have also been key players in the development of land management regimes for Lake District landscapes. Today around 40 per cent of the land in the Lake District is owned by the National Trust, the National Park Authority, United Utilities and the Forestry Commission and much of the remaining 60 per cent in private ownership is land that is farmed and managed subject to rigorous public policies and regulations, and with a high dependency on government grant-giving schemes. In practical land-management terms, the Lake District landscape is now for all intents and purposes "a national property".

An Inspirational 21st-century Landscape?

The question arises in these early years of the 21st century: how will these large and influential organisations exercise their responsibilities to manage the landscape or influence its management over the coming decades? Not surprisingly, the Lake District has a plan.

The plan for the future management of the Lake District has been prepared by the Lake District National Park Partnership, a grouping of representatives of some 25 public, charitable and private bodies with a stake in the management of the National Park. The overarching vision of this management plan (referred to as the "Partnership's Plan") to the year 2030 is to be:

An inspirational example of sustainable development in action. A place where its prosperous economy, world class visitor experiences and vibrant communities come together to sustain the spectacular landscape, its wildlife and cultural heritage.

This is not the place to elaborate on or analyse this lengthy and comprehensive document, but a few paragraphs are necessary to give a flavour of what it says about the landscape and how the views described later in this guide might evolve over coming decades as the plan is implemented.

The management plan proposes that the National Park should be an example nationally of "sustainable development in action" – no small aspiration. Sustainable development is usually defined as "meeting the needs of the present without compromising the ability of future generations to meet their own needs". Adopting this concept as the centrepiece immediately tells us that this plan is intended to be wide-ranging and inter-generational in scope, bringing in perspectives about the National Park's environment and our relation with it that go far beyond the simple notion of controlling development that could spoil the look of the landscape.

What is interesting is that many of the environmental challenges facing the Park that the management plan brings to the fore, perhaps for the first time in such a clear way, are not readily visible to the casual observer, and because of this are not yet apparent to many Lake District residents and visitors. An intrusive out-of-place building or structure is readily seen by everybody but poor river quality and low levels of biodiversity are not. Here are some revealing statistics from the National Park Authority for 2018: assessments undertaken in accordance with the requirements of the Water Framework Directive (see glossary) show that only 39 per cent of lakes, and the same percentage of rivers, had "high" or "good" ecological status. Some 20 per cent of all land in the Lake District is designated for statutory protection because of its biodiversity value (such as Sites of Special Scientific Interest), but 62 per cent of these sites are categorised as being in "unfavourable" condition. Out of twenty-two "priority habitats" (see glossary) in the Lake District, only eight habitats have an area in "favourable" condition that is greater than the area in "unfavourable" condition. Overall, the health of the Park's natural environment is presented as falling short of being an "inspirational example", but for most people the problems are largely out of sight. Then overlaying and exacerbating these local environmental challenges is climate change: higher temperatures, more extreme weather events (particularly flooding) and sea level changes. The devastation caused by Storm Desmond in December 2015 shook Cumbria to its core and thrust the dangers of climate change to the centre of thinking about land management in the long term.

Alongside these environmental issues, the management plan also highlights the continuing perilous financial state of the District's upland hill farms – farming businesses so crucial to maintaining what the Partnership calls the "harmonious beauty" of the Lake District and one of the key attributes of Outstanding Universal Value underpinning the Lake District's

The A591 road, a vital link between the north and south of the Lake District, was severed just south of Thirlmere during Storm Desmond in December 2015.

World Heritage status. The reality is that the produce from upland hill farms does not command sufficient financial returns in the market place to make them economically viable. In order to survive they need to earn income from other sources, including from government grants paid in return for undertaking work to improve the quality of the environment. The mantra for supporting UK farms in the future is "public payments for public goods", but exactly how this will work out as the UK reshapes its own independent agricultural support policies is uncertain.

A key question is whether the scale and pace of environmental change deemed necessary in the Lake District will be compatible with the traditional management of upland sheep farms and their distinctive landscapes. The management plan describes it as a "tension" between enhancing the natural environment and managing the cultural landscape, notably upland hill farming.

There is something a little tricky, to say the least, about seeking to sustain a landscape that has World Heritage status for its sheep farming traditions and landscapes, given that those landscapes are the product of constant evolutionary change in response to the imperatives of past times. A new imperative for us today, the management plan suggests, is to address ecological fragility in the context of climate change. Resolving this "tension" (of course, a bureaucratic euphemism for "a very fraught debate") is a real challenge that will test the Partnership. It is aiming to enhance the quality of the natural environment *and* sustain the viability of upland hill farming, but has given little indication about how this can be achieved in practice. The signs are that more holistic land-management planning approaches are to be pursued at the whole-valley scale, looking at the landscape as an ecosystem that provides "services" for human and environmental wellbeing such as food, wood, water, carbon management, regulation of floods, biodiversity, and non-material benefits such as recreational opportunities and spiritual well-being. Land-management choices are then about the mix of services required and how to deliver them.

What we don't yet know is whether the landscapes we are used to enjoying today are going to change materially or not. It seems inevitable that the new environmental imperatives will bring about changes – many more trees for a start – but how far will these go? Is there the prospect of the present sheep-farming landscapes gradually disappearing, notwithstanding World Heritage status? Or will mixed landscapes evolve in which sheep farming is less prominent; more bio-diverse but still farmed? Or could there be different approaches to landscape management in different Lakeland valleys? Or will the changes be so subtle and extend over such long time horizons that few will really notice? We don't yet know, although the Partnership's management plan does make it clear that the landscape will need to evolve as it always has.

This raises the question about how all of us, the wider public with an interest in the Lake District, will react to any changes. In thinking about our personal responses to the Lake District's landscapes, it is worth reflecting that William Wordsworth argued that the taste for romantic scenery is not with us at birth, but is something that must be gradually developed. Putting aside the slightly elitist undertones in his argument, the essential point is that cultural influences affect whether we "see" beauty in a landscape or not, and how we see it. Across the globe and within our own multicultural society there are wide differences in response to landscapes. Lake District landscapes do not inspire everybody. The way we see, understand and enjoy any landscape has been learnt from personal experiences and the cultural influences we have been immersed in during our lives. It is in effect, as Wordsworth suggested, a taste; but tastes can change. This begs the question: could we, if we had to, learn to see and love the landscape in other ways?

In the coming decades, it appears that our taste in landscapes may be challenged. We may find ourselves nudged away from our long-loved taste for a "beautiful" landscape and towards a new taste for a "healthy" landscape. Or to put it another way, we may be asked to see beauty in the landscape in a new way and learn to love it and find it just as inspirational. It could be that the management of the Lake District's cultural landscape is on the cusp of its next evolutionary turn; we'll have to wait and see.

Hows and Knotts: The Viewpoints

This book is a guide to 22 Lakeland views, but for the most complete experience an interesting view deserves to be matched with a viewing point that has a character of its own. This is what many Lake District hows and knotts obligingly provide.

Topographic features named "hows" and "knotts" on Ordnance Survey maps are common in the Lake District landscape and are concentrated here more than anywhere else in the UK. Their protruding presence is one of the ingredients that make up the landscape's character and provide an identity that is well recognised by visitors to the fells and adopted in the names of scores of farms, hotels, houses and streets throughout Cumbria. They are often well positioned to reveal the landscape interest and, with an average height of 350 metres, they are not so high that the details of a scene become lost or so low that the wider setting cannot be fully appreciated. Buildings, tracks, walls, crags, becks, pastures, woodland and the other ingredients of a scene can all be clearly identified and seen in context. But best of all, hows and knotts often provide an airy spot – land falling away beneath the viewer's feet in a way that adds to the whole experience of the view.

So, in praise of hows and knotts, a short biography follows. It is a story of their appearance, origins and names that takes us into the worlds of geomorphology, toponymy (the study of place names) and the early days of the Ordnance Survey.

All the hows and knotts within the Lake District World Heritage Site named on the Ordnance Survey's 1:25,000 maps were visited by the author over the course of a year. There were a number of objectives in mind: to identify viewing points from which interesting examples of Lake District cultural landscapes can be seen; to test the author's thesis that today's generally accepted definitions of hows and knotts are somewhat awry; and most importantly, to have fun visiting new territory. It was not a scientific survey. Of the 443 named hows and knotts that were found to be landforms rather than some kind of habitation, around 90 per cent actually felt the weight of the author's boots and the remainder were viewed from a short distance because of lack of public access or, occasionally, bone idleness.

Shapes

Big shapes in the landscape have already been looked at in this guide; now it is the turn of the smaller ones. It is perhaps necessary to say from the start that the words "how" and "knott" are not technical geological or geomorphological terms but common nouns adopted by communities over centuries to refer to landscape features and then recorded on Ordnance Survey maps and in other documents. Nevertheless, hows and knotts are physical landforms that are with us today as the result of geological and geomorphological processes over long periods of time. A mere half dozen hows and knotts appear to be depositional landforms, but for all intents and purposes it can be said that hows and knotts are rock-based erosional landforms.

Because "how" and "knott" are not technical terms, there are no technical definitions of their shapes, but what we do have are some descriptions based on the original meanings of the words:

> How: derived from the Old Norse word "haugr" that meant hill, mound, burial mound
>
> Knott: derived from the Old Norse word "knottr" (also "knutr") that meant knot, hard lump

These original meanings when applied to landforms imply, in general terms, compact shapes that are either smooth and mound-like ("hows") or craggy and knot-like ("knotts"). These are the descriptive definitions that are most commonly used today. Looking at the features in the field provided an opportunity to assess whether these descriptions adequately fitted with reality.

On the following pages are eight photographs illustrating the different shapes of hows and knotts together with some explanatory text. During the author's grand perambulation, it soon became apparent that two landform shapes kept recurring – what can be termed "mounds" and "steps". The mound shapes have a summit area with land falling away in all directions but not necessarily at the same gradient or over the same distance. The step shapes break the profile of sloping terrain with a levelling out, often close to horizontal, with a steep, craggy down-slope face. What was found in the field was that both hows and knotts can be either mounds or steps: 51 per cent of hows are mounds and 40 per cent steps; 42 per cent of knotts mounds and 56 per cent steps. (The missing percentages are where no protruding landform could be seen; this is discussed later.)

Examples of Shapes of Hows

Elf Howe, Kentmere (NY468001)
This can be considered a classic how displaying the key features ascribed to the Old Norse word haugr: a compact smooth-profiled mound without prominent protruding rock. However, only 32% of hows have an appearance similar to this.

Little Round How, Buttermere (NY208132)
This how could hardly be more different, a landform entirely of rock. The majority of hows display areas of rock, although the mix of rock and vegetation cover varies widely. The symmetry of the mound shape of hows also varies; in this example it has a distinctly asymmetric shape.

Looking How, Duddon (SD244999)
This how has a very different appearance: a step profile rather than a mound. The step is located part-way down the valley side and from the valley floor the profile is not readily apparent. This step profile is very common: 40% of all hows have a profile similar to this.

Eller How, Eskdale Moor (NY184034)
It is a curious fact that 9% of hows do not display any protruding shape at all; they are simply areas of sloping or undulating fellside. In these cases the name how is plainly not referring to an upstanding mound.

Examples of Shapes of Knotts

Pianet Knott, upper Eskdale (NY234046)
This can be considered a classic knott displaying the key features ascribed to the Old Norse word 'knottr', a compact and rocky knot-like feature. However, only 18% of knotts have an appearance similar to this.

Sharp Knott, Lamplugh Fell (NY107201)
This could hardly be more different, displaying the smooth, rock-free profile more readily associated with a classic how; 17% of knotts do not have any rocky features.

Piketoe Knott, Thirlmere (NY323168)
This is a knott displaying a step profile, very different from the classic rounded lump shape. This is the most common profile displayed by knotts – 56% of all knotts. There are no distinguishing differences between hows and knotts with step profiles.

Coldkeld Knotts, upper Eskdale (NY227079)
Generally both hows and knotts are compact individual features, but there are examples where several features are clustered tightly together and named in the plural. This is particularly the case for knotts: 17% of all knotts are in such cluster form.

With the original Norse meanings in mind, other aspects of the appearances of hows and knotts recorded were the prominence or otherwise of rocky features, the degree of compactness and the overall size. In terms of prominence of rock, 61 per cent of hows and 78 per cent of knotts display prominent areas of rock, with the remainder appearing as vegetation-covered features. In terms of size and compactness, around 65 per cent of hows are compact, medium-sized features (some 20 to 60 metres in height), while knotts display a wider range with 46 per cent that are particularly small or large or consist of a cluster.

The original question in the author's mind was whether the character of hows and knotts as observed today fits with their commonly accepted definitions. The conclusion must be that a definition made with the Old Norse words "haugr" and "knottr" in mind, that all hows are mound-like landforms and all knotts are rocky knot-like landforms, is plainly incorrect. Then during the grand perambulation, a second question arose: is it possible that there is actually no real difference between hows and knotts? Have the original, presumably different, topographical characteristics of hows and knotts been lost over time? What can be said is that hows are more likely to be a mound in shape, more compact and modest in size and more likely to have a smooth profile with complete vegetation cover. Knotts are more likely to be a step in shape, provide examples of particularly large landforms and more likely to display prominent rocky features. But most of these differences are marginal.

These conclusions make it difficult to give hows and knotts simple, distinguishing modern-day definitions that satisfactorily fit. This is considered further when discussing the naming of hows and knotts.

Locations

The map overleaf shows the distribution of all 443 hows and knotts shown on the Ordnance Survey 1:25,000 map that are landforms and not habitations.

Hows and knotts are rock-based landforms and their distribution in relation to the main geological groups shows that 63 per cent lie within the Borrowdale Volcanic Group areas, 15 per cent in Skiddaw Group areas and 16 per cent in Windermere Supergroup areas. The remaining 6 per cent are in the granite and other geological areas. (The main geological areas in the Lake District were discussed earlier in this book in the section on the physical landscape and are shown on the map in that section.) This predominance of hows and knotts in the volcanic area remains the case even when its larger geographic extent is taken into account with more than double the number of hows and knotts per square kilometre in the Borrowdale Volcanic Group area than elsewhere.

This concentration in the Borrowdale Volcanic Group areas can be largely explained by the immense variety of rock types: lavas, pyroclastic and volcaniclastic rocks. These rocks have different levels of hardness and structural character arising from their particular origins, making their ability to withstand erosion very variable. This variability exists over small

Within the map:

Cockermouth

Penrith

Pooley Bridge

Keswick

Bampton

Buttermere

Nether Wasdale

Ambleside

Ravenglass

Coniston

Kendal

N

Hows
Knotts
Haws
View point
World Heritage
Site boundary

0   km   10

Broughton

Newby Bridge

Distribution of Hows and Knotts (base map courtesy of LDNPA) distances and leads to profusions of hard outcrops standing out prominently from areas of less resistant rocks.

The four photographs on the pages following provide examples of landscape settings in which hows and knotts are found. They are most commonly located within the more heavily glaciated valleys – 60 per cent of hows and 65 per cent of knotts. Glaciated valleys are generally perceived as smooth troughs where intense glacial erosion has taken place, dramatically modifying the pre-Ice Age valley forms. That there are any rocky protrusions within them at all is perhaps a mild surprise. The majority (61 per cent) of the hows and knotts within glaciated valleys are on the valley flanks with the remainder on the valley floor or on the valley rim. Outside the heavily glaciated valleys, the remaining hows and knotts are found on top of and at the end of ridges and spurs, on a few fell summits and fairly widely in what can be termed open fell country.

Examples of locations of hows and knotts

Great How, Thirlmere (NY314187)
This is one of the 18% of hows that are located on the floors of glaciated valleys; knotts are
almost entirely absent from such locations. As its name implies, it is an unusually large how; only
11% reach such a size.

Gladstone Knott, Langdale (NY255045)
Some 10% of hows and 17% of knotts are located on the rim of deep glaciated valleys at the
sharp break of slope marking the boundary of glacial gouging. Many knotts at this position
appear to be "hanging" on the valley edge.

Lang How, Blea Rigg (NY318070)
Many ridges and spurs have hows and knotts on them in the form of mounds and steps. In this example, Lang How is a prominent mound on the ridge between Blea Rigg and Silver How.

Glade Haw, Duddon Bridge (SD201888)
A number of spurs and ridges terminate with a rounded "nose" named either a how or knott. 11% of all knotts are features of this kind.

Origins

It is a theme of this book that there is more enjoyment to be had from landscapes if we have some understanding of what we are seeing and why they are there. This equally applies to the hows and knotts we see in the landscape, but achieving an understanding of their origins is not so easy. Below are six diagrammatic profiles which summarise the earlier photographs and the profiles of hows and knotts when seen in the field.

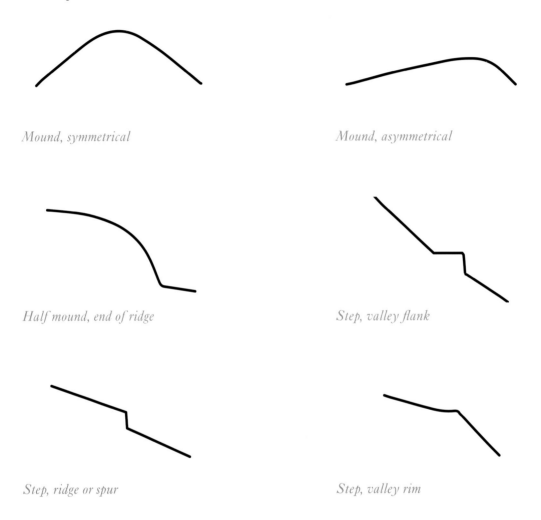

Mound, symmetrical

Mound, asymmetrical

Half mound, end of ridge

Step, valley flank

Step, ridge or spur

Step, valley rim

With varying degrees of fit, all hows and knotts are represented by one of the profiles and it would be nice to know why they appear as they do. As mentioned earlier, "how" and "knott" are not technical terms and have therefore not been the subject of specific study by geomorphologists. However, the work done by geomorphologists on erosional landforms in the Lake District and elsewhere provides a general idea of the processes involved in creating them and one of the profiles, the asymmetrical mound, is a hardy perennial in physical geography textbooks.

Clearly, in being protruding rock-based landforms, hows and knotts have particular characteristics that have enabled them to resist erosion to a greater degree than their immediately surrounding areas. The processes of erosion that were most important in shaping them were those occurring during the Ice Age, in particular erosion by moving ice, and during the immediate post-glacial weathering processes. Geomorphologists recognise two processes at work in the erosion and shaping of hard rock by moving ice: "abrasion" and "quarrying" (or "plucking"). Abrasion is a process whereby debris in the base of moving ice grinds out fine-grained rock particles, leaving the bedrock scoured and smoothed. Quarrying is the process by which a glacier can remove large fragments of rock by fracturing the bedrock and then carrying the broken rock away.

A good example of these processes at work at the individual landform scale is in the asymmetrical mound shape shown diagrammatically above and in the photograph of Little Round How earlier. This is a landform geomorphologists call a "roche moutonnée", literally meaning "sheep-like rock" in French (but see the glossary). Both hows and knotts can display this shape. This asymmetric landform has a gently rising, abraded up-ice side and a steep, quarried down-ice side and can vary enormously in size from a few metres to a few hundred metres in both height and length.

The surfaces and shape of a roche moutonnée reflect the distribution of stresses from overlying ice and water pressures on the rock protrusion as ice passes over it. As glacier ice rises up over the up-ice side of the rock protrusion, pressure becomes greater than normal in the basal layers of the ice. This lubricates passage over the protrusion and assists the downward migration of rock fragments to the bottom of the ice layer, increasing the effectiveness of abrasion. As the ice moves over the lee, down-ice side of the protrusion the pressures in the basal layers of the ice become much lower than normal, creating a cavity where the bedrock is fractured and pulled apart apparently in response to differential stresses placed on the bedrock from a combination of ice and fluctuating water pressures. The detailed structure of the bedrock will significantly affect both the up-ice abrading and down-ice quarrying actions and determine the final shape of the landform.

The asymmetric mound shape of the textbook roche moutonnée is best considered as just one example of the erosional effects of the processes of abrasion and quarrying. All shapes of hows and knotts are essentially the product of these two processes but with different end results depending on the nature and orientation of bedrock structures, nature and direction of the ice flow, pre-existing topography and other factors. However, information about how these factors shaped the different landforms displayed by hows and knotts does not appear to be readily available for the general reader. Nevertheless, the explanations given above do, hopefully, give a general idea.

Names

As already mentioned, "how" and "knott" are common nouns used by local people to identify the features they see every day. Most place names in England are compounds consisting of two "elements" with a specific element first and a generic element second, such as in Stonethwaite where "thwaite" is the generic element.

There is confidence amongst the experts that the place-name generic element "how" has its origins in the Old Norse word "haugr" and that the place-name element "knott" has its origins in the Old Norse word "knottr". However, this is not the same as saying that every place name containing "how" or "knott" comes from a common Norse source or meaning. The complication is that names change over time and the present use of the noun "how" or "knott" in a place name could be a substitute for an earlier name which has different origins, the result of regular sound changes in the speaking of a name or applied with a changed meaning to the name.

So the author's two conclusions, described earlier, that the appearance of many hows and knotts depart from the accepted original Old Norse meanings and that the physical differences between them are marginal is not too much of a surprise: the names are the product of a long evolution with communities speaking and applying the names in changing ways over time.

An intriguing aspect of this evolutionary point is that there are 24 named hows in the Lake District where it is not possible to identify a protruding landform at all. Often these are simply areas of gently sloping or undulating fellside. Why have communities endowed these places with the name "how"? A clue might be that the Old Norse word "haugr" also referred to burial mounds as well as simply mounds. The possibility that these tracts of land were so named because they were burial places or recognised by the Norse as ancient burial places is strengthened by their noticeable tendency to be close to known ancient sites (for example, Eller How close to the settlement on Eskdale Moor and Tongue How at the settlement on Town Bank, Kinniside). At these locations are we seeing place names handed down to us that reflect a meaning of "haugr" as burial place?

Pondering these questions leads to a more general question about why communities have given names to certain topographical features but not others. Why are some features named on Ordnance Survey maps whilst others, sometimes very prominent, are not? Putting this question to experts in place names and Ordnance Survey history generally prompts the same answer: "That's a good question!" The truth is we don't really know. We can only speculate that some topographical features may have been given names because they served or had served some practical purpose in identifying boundaries, meeting places, resources and such like.

When the Ordnance Survey first got to work, the collection and verification of place and topographic names was a major task alongside the early surveying work. The first mapping work covering the Lake District started in the mid 19th century and place names, including the names of topographical features, were recorded in Original Name Books as elsewhere in the country. Unfortunately, the majority of Original Name Books for the UK were lost during the bombing of London in 1940, but Cumbria is fortunate that some of the records for Cumberland, Westmorland and Lancashire have survived and are held in the National Archives at Kew. These records set out the authority for the names and spellings of places and features recorded on the map.

The Ordnance Survey's guidance to their first surveyors in England has not survived but the later *Instructions to Field Examiners* dated 1905 gives a good flavour of the practices followed. The Instructions state that place names should preferably be obtained from pre-existing printed or written sources, such as estate maps, valuation rolls and tithe maps. When local people are to be approached to provide the authority for names the Instructions set down a pecking order of the people in formal positions who should be approached, with landowners the first port of call, followed by postmasters, rate collectors and others, finishing with "gentlemen residing in the district". The Instructions then go on to say that "Respectable inhabitants of some position should be consulted. Small farmers and cottagers are not to be depended on, even for the names of the places they occupy, especially as to the spelling, but a well educated and intelligent occupier is, of course, a good authority". As an example, the Original Name Book for Bassenthwaite Parish in 1863 informs us that the authority for the name and spelling of Ling How "a rocky brow at the north end of Sandbeds in Bassenthwaite Common" comes from Mr J Thompson, Mr Joseph Graham and Mr Hudson Robinson, presumably people of some standing locally.

The Instructions then go on to deal in more detail with spelling and the compounding of names stating "Particular attention is required to the mode of spelling or compounding local names, and nothing shows an Examiner's intelligence or want of it more than the manner in which he attends to the spelling of names". The Ordnance Survey's conventions for the recording of compounded names are relevant to the naming of hows and knotts. The example is given that a name presented to the Examiner as "Parkhill" should be recorded as two words "Park Hill" if naming a hill, but "Parkhill" if referring to a farm on a hill. However, the Instructions also state that this convention should be followed only where local practice is unclear. Within the Lake District this convention is generally, but not always, followed with respect to names containing the element "how" or "knott". Where the element is contained in a single word, for example "Redhow", this invariably indicates a habitation of some kind; but in a departure from the convention there are 51 occasions where a two-word name, for example "Tom's How", was found to be the name of a habitation, not a landform. These habitations might originally have taken their names from an adjacent hill feature, but in many cases this is far from clear on the ground. A full list of hows and knotts that are landforms can be found at the end of this guide.

A final point on spelling is the spelling of "howe" and "haw" rather than "how". All three spellings share origins with the Old Norse "haugr" and are therefore treated the same in this guide. "Haw" is almost exclusively found in the south-western parts of the Lake District, presumably reflecting local pronunciation. Why an "e" is added to the end of "how" in some cases is unclear and appears to have been the subject of some differences of opinion. In the Ordnance Survey's Object Name Book for the 1911–12 revision for Westmorland, there is a note against two revised entries where the "e" has been deleted stating that "The authorities object to the 'e' on Howe". Which authorities and why is not explained.

The Views

Introductory Guidance

Over the following pages, 22 views are described from prominent viewpoints on hows and knotts in different parts of the Lake District World Heritage Site. Earlier in this guide there was a general overview of the making of the Lake District's physical and human landscapes; in what follows, local detail and examples are presented. Each view description points out any notable features in the physical landscape followed by points of interest relating to human activity in shaping the landscape legacy we enjoy today. By the time you have absorbed them all, you will have traversed much of Lakeland's cultural landscape.

In making this journey you have a choice about your level of exertion. For the purists, each view must be appreciated by standing on the how or knott, even though on occasions some rough ground may need to be crossed. For old hands who are familiar with the territory (and other similarly relaxed readers) this guide, the Ordnance Survey map and the panoramas on the book's website, all viewed from the comfort of an armchair, should provide the right mix of mental exertion and opportunities for reminiscing.

Warning. For those venturing out please take note that this book is *not* a walks guide. To visit the viewpoints you need to be familiar with walking in the fells, be able to work out your own route from the OS 1:25,000 map and know how to navigate. Guidance on approaches to the viewpoints is given in each description. Some viewpoints involve traversing rough ground. You will need to be properly clothed and equipped for the weather and other conditions – expected and unexpected. If you are unsure what this involves please seek experienced advice.

All the viewpoints are understood to be open to public access, but in the event of any difficulty, retreat gracefully and report the problem to the National Park Authority.

For those readers viewing from an armchair there are panoramas of every view on the book's website, www.lakelandviews.uk. They are best viewed on a tablet; expand the image and then move around the view at will.

Please note that the location references in the following pages are of the viewpoint, which in some cases is a short distance from the high point of the how or knott.

Allen Knott, Windermere

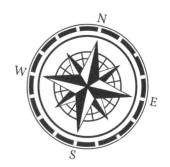

An easily accessible but rarely visited open viewpoint that provides a wonderful view of Windermere and some idyllic wooded and pastoral scenes

Location:	NY414011; N54°24.069 W002°54.283
Height:	220 metres
Geology:	Windermere Supergroup
Ownership:	National Trust
Access:	Open access land; readily accessible from the footpath off Moorhowe Road
Website panorama:	From north through west to south

Allen Knott from the east

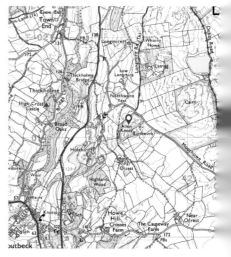

Location of Allen Knott

Q *The View*

This superb view over the northern reaches of Windermere provides a contrast between the subdued terrain of the lake shores and the more distant, dramatic heights of the Langdale Pikes and other central fells. With the towns of Ambleside, Windermere and Bowness just out of sight, Windermere's eastern shore, below, is a peaceful scene of woodland, parkland and pastoral fields interspersed with partially hidden grand houses built by wealthy outsiders in the 19th century. Beyond, on the lake's sparsely inhabited western shore, the well-wooded low fells of Claife Heights reflect a different history of woodland exploitation followed by ambitious woodland replanting for amenity and commercial purposes. Today these northern reaches of Windermere are a popular location for recreational boating activities controlled through a controversial speed limit.

View from Allen Knott looking south-west

Your viewing position: Looking west across the lake

Windermere's north basin: Before you is Windermere, lying in a narrow trough gouged out of the relatively soft rocks of the Windermere Supergroup by glaciers moving southwards, right to left, off the high volcanic heights of the central Lake District. The lake is over 17 km long, divided into distinct north and south basins, with its maximum depth, 64 metres, in the north basin in front of you. The basin was formed by the glaciers within the Trout Beck, Stock Ghyll, Scandale, Rothay and Brathay valleys converging and excavating the sedimentary rocks to a depth, by some estimates, of at least 100 metres below sea-level. As the valley glacier retreated, water was left impounded to create the lake, and a considerable quantity of glacial and alluvial material was subsequently deposited in and around the over-deepened rock basin. Close by, to your right just beyond Trout Beck, glacial till, deposited and shaped in the form of drumlins, can be seen amongst the trees.

Allen Knott hillfort: Allen Knott is the site of a scheduled ancient monument, described as an example of a "slight univallate hillfort", a fort that is typically fairly small, situated on a hilltop and defined by a single line of earthworks. The earthworks can be seen on the northern and western sides (you probably crossed them when approaching Allen Knott from the public footpath) but the southern and eastern parts of the site have been lost to agriculture and quarrying. The site probably has origins in the late Bronze Age or early Iron Age. There has been some discussion amongst historians about the possible later function of this prominent site during the Roman period because of the site's proximity to two important Roman roads – the southern end of High Street, which runs south from near Penrith, and the road between the Roman forts at Watercrook, in Kendal, and Galava, in Ambleside. Although Allen Knott is not thought to be at the point of intersection, evidence of the Watercrook to Galava road has been found in several places, suggesting a line running immediately below where you are sitting.

Villas and designed landscapes: In the late 18th and throughout the 19th century, the shores of Windermere, particularly the east shore below you, were a much-sought-after location where wealthy industrialists and professionals from Lancashire, Yorkshire and beyond built their grand country houses (commonly termed "villas"). These houses with their designed landscape gardens, containing water features, Picturesque tree planting and arboretums of exotic pines and conifers, were built within a landscape displaying the sought-after Arcadian beauties of the Lake District. Immediately below you is Holehird (a Grade II listed building with gardens by the famous Lancashire-born landscape and garden designer Thomas Mawson); close to the shore are Hodgehowe, Calgarth Park (Grade II), Cragwood (with gardens by Mawson), Brockhole (a Grade II historic park and garden, also by Mawson), Langdale Chase (Grade II); and across the lake the castellated Wray Castle (Grade II*) and Brathay Hall (Grade II). These grand residences provided an opportunity for the owners, their family and friends to enjoy the fashionable beauties of the Lake District, and for the owners to demonstrate to others in their strata of society their wealth and sense of good taste.

Not surprisingly, a proposal in 1876 to extend the railway from Windermere to Ambleside (and possibly on to Keswick) through this landscape with a viaduct over Trout Beck caused outrage. It was firmly resisted by the influential villa owners supported, amongst others, by John Ruskin who displayed his worries about easier rail access for working-class passengers with the words "I don't want them to see Helvellyn while they are drunk". This anti-railway

Wray Castle, built in 1840 for James Dawson, a Liverpool surgeon

campaign was one of the important staging posts in the development of the Lake District conservation movement. Whether in the 21st century we would be bemoaning a Victorian legacy of low-carbon transportation along the length of the busy A591 corridor is perhaps debatable!

Troutbeck village: To your right is the conspicuous, east-facing linear village of Troutbeck, sited on steep fellside well above Trout Beck. This architecturally superb village with origins in the medieval period is a designated conservation area. Its linear layout follows a spring line with clearly separated groups of farms and buildings (sometimes referred to as "bye-hamlets") stretching from Townend at the south end to Town Head at the north. The linear arrangement enabled each farmer to easily access their lower land in the valley bottom and the upper grazing commons. During the period of substantial rebuilding in stone during the 16th and 17th centuries there were up to fifty "statesmen" farmers in the village, dominating community and farming life. One of these farming families, the Brownes, lived at Townend, now a Grade I listed building in the care of the National Trust. There are a grand total of 26 listed buildings in the village, which makes a wander amongst them a real treat for those with an interest in vernacular architecture. The present field boundaries, immediately upslope of the road and between the road and Trout Beck, reflect the medieval pattern of inbye and intake fields. The common pastures were divided into three fenced and gated areas, known as "hundreds" (named on the Ordnance Survey map, north-west of the village), where cattle were kept, except during winter when they were brought down after hay and other crops had been taken in. The Troutbeck Hundreds were enclosed by Act of Parliament in 1831, introducing the typical rectangular field patterns into the landscape.

West shore tree planting: In the early 18th century the landscape opposite you on Windermere's west shore was largely denuded of trees because of their exploitation over many centuries to provide charcoal for the iron-making industries of High Furness. Now it is one of the most densely forested tracts of landscape in the Lake District. This is the legacy of extensive planting that got under way in the late 18th century, prompted by a new commercial impetus, overlain by inspiration from the Picturesque movement to improve landscapes where opportunities arose. One of the prime movers was John Curwen (the owner of Belle Isle, which lies just off Bowness, way to your left). In the late-18th century, Curwen bought up large tracts of land on the west shore to plant trees to supply timber for the growing industries and towns of west Cumbria where he had mining and other business interests. He planted a huge number of trees on the slopes of Windermere with a mixture of both commercial and aesthetic objectives. The woodlands produced timber, largely larch, spruce and pine, and coppice wood (oak, ash and hazel) for charcoal and bark for the tanning industry. Today the lower slopes of mixed deciduous woodland, including small tracts of ancient semi-natural woodland (see glossary), are managed by the National Trust largely for amenity purposes, while the higher land is managed by the Forestry Commission.

Windermere's water: The chances are that you are standing on Allen Knott on a nice day when Windermere looks serene. However, all is not quite as it seems: the water you are looking at hosts a number of problems that present some challenging water-management issues. For much of the 21st century the quality of the water has declined, with high phosphate levels arising from domestic effluent and agricultural run-off. The absence of a public mains sewerage system to serve the area between Waterhead at the head of the lake and Windermere/Bowness contributes to the problem. The high phosphate levels have stimulated algal growth, which has reduced light and oxygen levels for invertebrates and fish, as well as spoiling recreational activity. Soil erosion on agricultural land within the lake catchment has led to sediment run-off that is carried down by rivers and adds to the problems by silting fish spawning grounds, such as those for the rare Arctic Char. Lake banks and shores have been extensively damaged by human activity, leading to the loss of most of the lake's reed beds. Several native plants and animals are declining whilst other, non-native, invasive species are thriving. Overlying these problems is the growing impact of climate change, notably the damaging fluctuations in lake water levels arising from floods and droughts affecting lake-shore businesses, boating activities and wildlife.

All these issues are now being tackled through multi-agency improvement programmes for the whole Windermere catchment and some of them, such as phosphorus levels, are now being turned around following infrastructure investment. Local communities, businesses, farmers and visitors are all being engaged to make their own contributions, such as flood mitigation and resilience planning, increasing woodland cover and reducing the use of polluting fertilisers and household products. It is anticipated that getting Windermere back into a really healthy ecological condition will take several decades.

"Quiet enjoyment": The north basin of Windermere in front of you has long been a prime location for boating activity in the Lake District. Prior to 2005 this would have been where fast power boats and water skiing were a major part of the lake scene. Now there is a speed limit of 10 knots (11.5 mph) in place. The introduction of this speed limit was one of the most controversial decisions in the history of the National Park, and the upset it caused has still not died down. As the number of fast power boats using the lake increased considerably towards the end of the 20th century, the National Park Authority and others became concerned about the impact of noise levels on local residents and wildlife, the risk of accidents and the damage to shoreline habitats. Water skiing and other water sports organisations argued that any problems could be overcome by improved management of the water space, through zoning different activities and various other measures. The National Park Authority judged this impractical and, following a public inquiry, the Government in 2000 approved a 10 mph speed limit (subsequently amended to 10 knots) that came into effect in 2005.

Wakeboarding on Windermere

In reality the controversy was not just about the specific case for and against fast power boats on Windermere, it was also about what kind of national park we want. Back in 1991, a National Parks Review Panel recommended to the Government an amendment to the statutory purposes of national parks, to include the phrase "To promote the quiet enjoyment and understanding of the area", and further recommended that the forms of outdoor recreation encouraged should *only* be those that involve the quiet enjoyment of the parks. When the subsequent Environment Bill was going through its parliamentary stages, the references to "quiet enjoyment" were deleted, and the phrase is not included in national parks' statutory purposes in the Environment Act 1995. What the Act does require from every national park, however, is a statement of the area's "special qualities". One of the defined special qualities for the Lake District is "Opportunities for Quiet Recreation" – a phrase not commanding the same weight as the wording originally recommended to Government. Whether the Lake District should be managed as a place for quiet recreation only or whether adventurous, if sometimes noisy, recreational activities can also have their place continues to be contentious.

Bell Knott, Patterdale

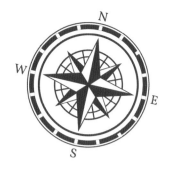

A spectacular viewpoint closely surrounded by dramatic craggy fells, overlooking Brothers Water and the uppermost reaches of Patterdale

Location:	NY394109; N54°29.421 W002°56.152
Height:	450 metres
Geology:	Borrowdale Volcanic Group
Ownership:	National Trust
Access:	Open access land; readily accessible on the footpath traversing High Hartsop Dodd
Website panorama:	From west through north to south

Bell Knott, centre skyline, from the north-east near Caudale Bridge

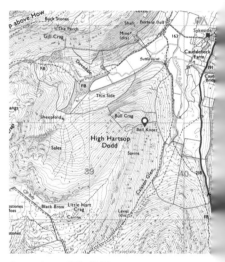

Location of Bell Knott

📍 The View

This is an excellent bird's-eye view of the valley floor with its interesting evidence of the area's varied history. There was a transient human presence during Neolithic and Roman times, but the known story really starts in the medieval period with features on the ground revealing how the lord of the manor and his tenants managed the land. Mining and smelting lead became a part of the local economy from the late medieval period onwards, reaching a peak in the 19th century when this would have been a busy and industrious scene. Today this upper part of the valley is very much quieter, engaged in sheep farming and providing for visitors at the Sykeside camp site and inn below.

View from Bell Knott looking north

Your viewing position: Looking north, down-valley, over Brothers Water

Brothers Water: This small lake is some 16 metres deep and lies in a glacially eroded rock basin that, during the immediate post-glacial period, extended further south towards you before being filled in with glacial and alluvial sediment. The shape of the lake is curious, but largely natural despite appearances to the contrary. The straight western shore is aligned to a geological fault and the northern shore is at the southern boundary of the alluvial fan laid down by Hayeswater Gill, where it enters the main valley from the east. The fan has also directed the Goldrill Beck outflow from Brothers Water over to the far, western side of the valley floor. The lake is a Site of Special Scientific Interest (SSSI) especially important for its range of bottom-rooted plant communities and animals dependent on them. Extending southwards on the southern shore there is a good example of habitat gradation: open water, lake edge gravel, birch and willow carr, acidic marshy grassland and neutral grassland with scattered trees. The most recent survey (in 2010) of the condition of the SSSI by Natural England assessed it to be "unfavourable, declining", particularly on account of the presence of non-native invasive species – the North American Western Waterweed. A university survey in 2016 found very high concentrations of lead in the lake sediments, the legacy of past lead mining nearby which is described below.

Early human presence: The earliest evidence of human presence in the valley is some Neolithic rock art found near Beckstones Farm, north of Brothers Water. In front of you at the toe of High Hartsop Dodd just beyond the barn there are the vague outlines of an ancient settlement. This is a Romano-British settlement, possibly from the 2nd century, which comprises an enclosure bank containing five hut circles. The site is a scheduled ancient monument. There is an opinion, not accepted by all experts, that there was a Roman road that came south along the western shore of Brothers Water and ran just west of Kirkstone Beck all the way over the Kirkstone Pass. The postulated line can be readily seen from the road leading up to the pass.

The farming landscape: As owners of much of land before you, the National Trust has undertaken detailed research to build a history of agricultural land-use and settlement from the Norman period onwards. From where you are sitting, you have a bird's eye view of the evidence of this history. Following the arrival of the Normans in Cumbria in the late 11th century, land was allotted to Norman lords and this valley land became the Manor of Hartsop. With a rising population in the 12th and 13th centuries, communal farming by the lord's tenants was organised around valley-bottom arable and meadow land enclosed within a walled ring garth. The probable line of this ring garth ran from the south-west corner of Brothers Water towards you following the edge of the valley floor, round into Dovedale as far as Hogget Gill, then skirted around the lowest edge of High Hartsop Dodd, below you, and then southwards for 500 metres before crossing Kirkstone Beck, and then taking a wavy course north past Caudalebeck Farm and Sykeside to the south-east corner of Brothers Water. Within the ring garth was the common open field, High Hartsop Field, used as meadow land and for growing arable crops. On the east side of the ring garth, a settlement of tenants developed in the vicinity of Caudalebeck Farm, known as High Hartsop. On the west side, the lord retained the lands in Dovedale and west of Kirkstone Beck and Caiston Beck as demesne lands; this was where his hall was built. To the north of Brothers Water, another settlement developed in the side valley of Hayeswater Gill known as Low Hartsop, today called simply Hartsop.

At some date after the ring garth was built, the demesne land at the head of Dovedale was enclosed, including the land left and below where you are sitting, called Thin Side. This enclosed land was managed as a vaccary, but little further enclosure took place on this demesne land until after 1600. Further east, however, around High Hartsop, intaking of land by tenants for cultivation and cattle pastures led to enclosures between the ring garth and the line of the present-day road. Then later in the 17th century there was further intaking on the east side of the road, south of Caudale Bridge and northwards up to Brothers Water. These latter walls display a markedly rectangular layout and are largely constructed of quarried stone rather than the field stone of earlier walls. By this time farming was increasingly dominated by sheep rearing alongside food crop cultivation. Further enclosures followed in the early 18th century, including the walls at and below Bell Knott, on the lower slopes of Middle Dodd and the long wall running south along the east side of the road from Caudale Bridge. By the end of the 18th century, widespread changes in the farming economy were leading to farm amalgamations and abandonment; High Hartsop had ceased to exist as a community and the enclosure of common land was getting under way. Parliamentary Enclosures from the 1860s through to the 1890s resulted in land being enclosed and allotted to new owners at High Hartsop Field and on the higher fells. Although largely out of your view, the straight line walls and rectangular field patterns typical of this time can be seen on the Ordnance Survey map on Hartsop Dodd and further west over to High Street.

Hartsop Hall: In front of you, below Wood Side, is Hartsop Hall, a Grade I listed building owned by the National Trust. The original house, built for the Lord of the Manor, appears to have origins in the 14th century, but the farmhouse it is today was built largely in the 16th century, with 17th- and 18th-century additions. From 1638 until the 20th century the house was part of the Lowther family estate, but it appears that for most of the time it was rented out to tenants. Hartsop Hall and the rest of the estate, including Brothers Water, passed into the National Trust's possession in 1947 after the estate was given in lieu of death duties to the Treasury, which then offered it to the Trust as an unconditional gift; the first time this had been done.

Hartsop Hall

Lead mining: It is hard to view this scene as a past industrial centre, but in its own small way it was. Beside Hogget Gill, in Dovedale down to your left, there is evidence of a late medieval iron bloomery, which would have used charcoal from local woodlands. The site was later used for a lead smelting mill in the 17th and early 18th centuries, with the local woodland producing "white coal" (chopped wood dried over a fire) to fuel the smelting hearth and a water wheel powering the bellows. The lead was mined across the valley at the Hartsop Hall Lead Mine; remnants of the site can be seen to the left of Hartsop Hall, a short distance up the fellside. This mine was most productive during the 19th century and was powered by water taken from the upper part of Dovedale Beck via a leat which is still identifiable today. The mine finally closed in 1942.

Low Wood SSSI: The prominent woodland in front of you on the flanks of Hartsop above How – comprising Low Wood, north of Brothers Water, and Wood Side to the south – is ancient semi-natural woodland. It is designated a Site of Special Scientific Interest and is also part of the internationally designated Ullswater Oakwoods Special Area of Conservation. During the medieval period, these woods were part of the demesne lands, although tenants probably had rights to collect wood and timber from certain areas. In later centuries, Wood Side and the woods in Dovedale, unlike Low Wood, appear to have been seriously depleted for the production of white coal and charcoal. Evidence of fifteen charcoal burning platforms has been found in the lower part of Woodside and twelve in Dovedale. By the early 20th century, however, Wood Side was showing signs of recovery that has continued although its character differs from Low Wood, reflecting the difference in past management. Today the SSSI is valued for the quality and range of tree types present, in particular well-developed stands of sessile oak and ash-hazel woodland. The last survey of the woodlands by Natural England (2012) summarised their condition to be "unfavourable, recovering" largely because of patchy natural regeneration. The Ullswater Oakwoods, including these woodlands, are subject to a Natural England Site Improvement Plan being implemented in association with the landowner (see the pages on Bell Knott, Ullswater).

Water management: Major floods in recent years have meant that the management of water run-off in Lake District valleys has now become a very live issue. The valley land below you provides a small example of the legacy of past water management that has to be lived with and planned around today. Over time, changes have been made to the natural courses of Caudale Beck and Kirkstone Beck on the valley floor. Caudale Beck originally flowed north to Brothers Water after arriving in the main valley, but at some point it was diverted to keep it on its westerly course until it reached Kirkstone Beck. This possibly took place in the late-medieval period as part of new intaking of land in this vicinity. The present course of Kirkstone Beck has been fixed by stone pitching and channelisation over an extended period. In the 19th century its course, just before entering Brothers Water, was shifted slightly east to its current position, presumably as part of the scheme of enclosure and drainage of the former common field. In the late 19th century, two new drainage channels were cut nearby to take water into Brothers Water, but despite these efforts the land quality remained poor for agricultural use as can be seen today. The aim of these drainage works and their regular maintenance was, as one historical document put it, "to afford a free discharge of flood waters"; an objective replaced today with the new mantra of "slow the flow".

Channelisation of Kirkstone Beck, south of Brothers Water

Hartsop above How: Finally, a curious matter: the naming of the long ridge in front of you. Alfred Wainwright in his guide to the Eastern Fells queries the hill name "Hartsop above How". So too did the present author when he saw the names "Patterdale below How" and "Hartsop above How" on the 1863 edition of the Ordnance Survey six-inch map – clearly the administrative names of two townships within Patterdale parish. Ordnance Survey staff kindly agreed to investigate and revealed that when the Old Series one-inch map (published in 1867) was prepared, the township name "Hartsop above How" was placed across the ridge in order to avoid the name being split over two adjacent map sheets. This had the unfortunate effect of suggesting to the map reader that this was the name of the ridge. In 1976, an Ordnance Survey review formally accepted "Hartsop above How" as the hill name on the grounds that everyone was using it – but everyone was using it only because the Ordnance Survey map said so!

Bell Knott, Ullswater

A fine viewpoint overlooking Ullswater's upper and middle
reaches and with views towards Patterdale's enclosing fells

Location:	NY379193; N54°33.930 W002°57.675
Height:	450 metres
Geology:	Borrowdale Volcanic Group
Ownership:	National Trust
Access:	Open access land; easily accessible from adjacent public paths
Website panorama:	From north through east to south

*Bell Knott, centre just below skyline,
from Glencoyne Bay*

Location of Bell Knott

Q The View

This is a view dominated by Ullswater. The varied surroundings of the lake include
unpopulated precipitous fellside with deciduous woodland and shorelines that have been
the setting for scattered early settlement, medieval deer parks, 18th- and 19th-century
Picturesque parkland and a famous association with William Wordsworth. The lake, with

its location close to the popular summit of Helvellyn, has become one of the Lake District's premier tourist destinations based on the 19th-century lead-mining village of Glenridding.

View from Bell Knott looking north-east

Your viewing position: Looking directly across the lake towards Birk Fell

Serpentine Ullswater: One of the charms of Ullswater, whether on the water or walking along the shores, is its changes in direction, quite unlike any other lake in the District. This distinctive shape is the combined result of underlying geology and the erosive power of the valley's glacier during the Ice Age. The rocks underlying where you are sitting, above you and rightwards around the head of Ullswater to Place Fell and Birk Fell are lavas and tuffs of the Borrowdale Volcanic Group. In contrast to these steep and craggy fells are the gentle slopes below you and to your left, leading down to the shoreline and across the lake on the far shore by Sandwick. This more gentle terrain is underlain by older Skiddaw Group rocks consisting of relatively soft shales and mudstones. Major folding and faulting divided these rock sequences into segments, separated by multi-directional lines of weakness, which were exploited by later glacial erosion resulting in Ullswater's changes in direction. A series of glaciers moved north-east off the Helvellyn range, converged in Patterdale and eroded along a north-south fault to create the short upper reach to your right, and then followed the east-north-east alignment of another major fault – the Ullswater Fault – to create the long middle reach to your left. Then another fault, the Howtown Fault, explains the kink between the middle and lower reaches. In the post-glacial period, powerful debris-laden becks have deposited alluvial material in the lake creating distinctive arcuate deltas at Aira Beck to your left, Glencoyne Beck below you and the largest at Glenridding Beck just visible to your right. During Storm Desmond in December 2015, Glenridding Beck caused serious flooding in the village of Glenridding and, in a graphic demonstration of how delta formation works, deposited vast quantities of boulders on top of the delta land occupied by the steamer pier, access road and other buildings.

Early settlement: The nature and dates of early settlement in the Ullswater valley are unclear, sites in all probability being occupied, abandoned and then resettled over time. Suitable land was generally in short supply, but the shore below you and to your left provided an obvious opportunity. The date for the settlement remains down to your left (marked on the Ordnance Survey map) is uncertain: possibly Iron Age, but probably better viewed as multi-period. Official records describe it as a settlement of four hut circles with an enclosing bank. In the immediate vicinity there is evidence of numerous clearance cairns, land drainage and agricultural works. Firm evidence of later pre-Norman settlement here is largely absent.

Gowbarrow Parks: The low fellside below you and to your left is named on the Ordnance Survey map as Glencoyne Park, Gowbarrow Park and adjacent, out of sight, Swinburn's Park. From medieval times, when it was part of the Manor of Greystoke, until the early 19th century most of this land was used as a deer park. Deer parks originated as symbols of medieval lordly power and wealth, and functioned as game reserves. Wild deer would be encouraged to enter over irreversible deer leaps and provided both sport and meat. Over time many deer parks became largely ornamental with the deer part of an attractive, designed landscape; however, in this case a record as late as 1794 refers to the presence of 600 fallow deer, suggesting the original function lived on longer here than elsewhere. A map of 1800 by the Keswick tourism entrepreneur Peter Crosthwaite (see pages on Brown Knotts) refers to the whole area from Gowbarrow House to Glencoyne Bay as "Gowbarrow Parks" and the long, walled boundary depicted on his map is still largely intact today.

Glencoyne: One of the finest reminders of the Lake District's period of statesmen farmers can be seen down to your right on the edge of the Glencoyne Beck delta. The buildings at Glencoyne Farm were built in the early 17th century during the period of widespread rebuilding in stone. With their distinctive huge circular chimneys they are some of the finest vernacular farm buildings in the District, now protected as Grade II* listed buildings. The farm is today a notable Herdwick farm with land in the Glencoyne valley, Glencoyne Park and on the open fell. The whole farm, including the land where you are sitting, was gifted to the National Trust in 1948 by the family of Sir Samuel Scott, an active National Trust committee member. A short distance up the valley, just visible in the trees below Glencoyne Wood, is a row of houses curiously named Seldom Seen. They were built, together with a school, around 1839 as cottages for miners who worked at the Greenside lead mine in the adjacent valley and their families.

Picturesque appreciation: By the late 18th century, the fascination with the Picturesque began to impact on the increasingly appreciated scenic qualities of Ullswater. The Howard family had succeeded to the Greystoke estate in 1571 and around 1795 Charles Howard, the 11th Duke of Norfolk, built a hunting lodge, Lyulph's Tower, which can be glimpsed to your left just inland of the far side of the Aira Beck delta. It was built in a Romantic Gothic castellated style and was one of the first residences in the Lake District to be built with the view and landscape setting firmly in mind. During the 19th century, Gowbarrow Park and the land around Aira Force were landscaped by the Howard family as Picturesque parkland with tree planting, paths and bridges, and in 1846 an arboretum was planted below the force with over 200 specimen conifers. Prompted by fears of development, the National Trust bought much of Gowbarrow Fell, including the environs of Aira Force, in 1906 through public subscription.

Glencoyne Farm

Ullswater attracted the early Picturesque visitors with the poet Thomas Grey visiting in 1769 and Thomas West a few years later, both identifying several viewing stations close to the lake's shore. William Wordsworth enjoyed Ullswater enormously and would walk over from his home in Grasmere to visit, and even planned to build a house on land he bought in Patterdale. The land already had a cottage called Brow How (now known as Wordsworth Cottage) but his new-build project did not come to fruition. On 15th April 1802 William and his sister Dorothy were walking in woods in the vicinity of Glencoyne Bay and came upon a display of daffodils so entrancing that Dorothy recorded the scene in her journal. The encounter and Dorothy's charming description subsequently inspired William's most famous poem, 'I Wandered Lonely as a Cloud', commonly referred to as 'Daffodils'.

In 1846 the railway station at Penrith opened and tourists were able to visit the Ullswater valley more easily. Hotels and lodging houses opened in Glenridding and Patterdale and a Romantic and Victorian landscape developed along the northern shore. However, the initial growth of Glenridding owes rather more to the Greenside Mining Company that opened up a large and very productive lead mine further up the Glenridding Beck in the 1820s. A large number of miners' houses were built in Glenridding, and between 1821 and 1841 the population of the parish of Patterdale doubled. With the mine complex, huge though it was, tucked up the Glenridding Beck, valley tourism and mining seem nevertheless to have got along side-by-side. The mine did not close until 1961.

Woodland management: Woodlands are very prominent in your view and most of them are ecologically important habitats, with some designated Sites of Special Scientific Interest (SSSI). The wooded areas on the northern shores of Ullswater are largely owned by the National Trust and the rest are privately owned. These owners are incentivised through publicly funded woodland grant schemes to manage their woodlands to enhance biodiversity, mitigate climate change and improve public access, alongside the production of commercial wood products. The woodland in Glencoyne Park, below you and left as far as the A5091, although not identified as a SSSI, is an important type of open woodland habitat known as "wood-pasture". This land is grazed by cattle and valued for its trees, especially veteran and ancient trees, and the plants and animals that they support. Further left, beyond the Aira Beck delta on the lower slopes of Gowbarrow Fell, there is ancient semi-natural woodland with some very fine stands of alder and wych-elm. Much of this woodland is designated SSSI.

Then, turning to your right, there is Glencoyne Wood: the higher slopes are ancient replanted woodland and the lower slopes ancient semi-natural woodland where sessile oak is the dominant canopy tree. These lower slopes are designated a SSSI and are part of the internationally designated Ullswater Oakwoods Special Area of Conservation (which includes Low Wood, adjacent to Brothers Water; see the pages on Bell Knott, Patterdale). When last surveyed by Natural England (in 2010), this SSSI was rated "unfavourable, recovering". The wood is managed in line with Natural England's Site Improvement Plan for the Ullswater Oakwoods that aims to reduce damaging grazing and the impact of non-native tree species within the wood.

Finally, looking across the lake opposite you, the slopes of Birk Fell and Place Fell are designated a SSSI with part-ancient semi-natural woodland, including the most extensive stand of juniper in the Lake District. At the last survey undertaken by Natural England (in 2012), this SSSI as a whole was recorded as being in "unfavourable" condition, but with parts of the site "recovering" where there is some natural regeneration in the absence of grazing deer.

Water management: Given the view in front of you, it is perhaps not surprising that the future management of the water environment throughout the Ullswater valley is an important local priority. The lake is subject to a protective management regime by virtue of being part of the internationally important River Eden Special Area of Conservation and is also subject to the requirements of the EU Water Framework Directive that seeks to achieve "good status" for all ground and surface waters: that is, good-quality water and water-dependant ecosystems within the catchment. Currently, the quality standards are just about being met, but Natural England and the Environment Agency have concerns about the levels of phosphorus and the risk of eutrophication (nutrient enrichment) of the lake waters. The sources of pollution arise from sewage disposal facilities that cannot fully handle the tenfold increase in population in the Ullswater catchment during the summer months and from the run-off of pollutants from farms and agricultural land. The agencies work continuously with landowners to reduce the agriculture and sewage discharges that cause these problems. In addition, there are lead concentrations in the lake sediments, particularly at the south end of the lake, assumed to originate from the old Greenside Mine.

Overlying these problems is the urgent local issue of flooding following the devastating impact of Storm Desmond on the valley in December 2015. Glenridding was severely flooded and damaged and at the foot of Ullswater the fine, old bridge at Pooley Bridge was unceremoniously swept away. Before the arrival of Storm Desmond the communities living in the Ullswater valley were participating in a pilot "whole valley" management plan as part of the work in implementing the Lake District National Park Partnership's Plan. The plan has stimulated an inter-agency and community engagement process that brings together the challenges of valley water management and other community social and economic initiatives. The community of Glenridding now has a Flood Action Plan.

Flooding at Glenridding after Storm Desmond in December 2015

Boat How, Ennerdale

A good viewpoint on the south side of Ennerdale Water in the valley hosting the experimental "Wild Ennerdale" project

Location:	NY111137; N54°30.619 W003°22.459
Height:	363 metres
Geology:	Ennerdale granite
Ownership:	National Trust
Access:	Open access land; access from the bridleway that terminates west of Boat How; make for the junction of four walls beyond the end of the bridleway and then traverse over rough ground
Website panorama:	From east through north to north-west

Boat How from the north-west

Location of Boat How

◉ The View

Here is a valley scene of unusual isolation and wildness that hosts the experimental multi-agency "Wild Ennerdale" project. Low-level human settlement through history has allowed archaeological evidence from the Bronze Age to the present to remain relatively undisturbed. Even today this large valley has no public road and only a small resident population: the only glimpses of permanent settlement are to the west where the valley opens out into the coastal plain. The scene is dominated by Ennerdale Water surrounded by a mixture of commercial forestry, deciduous woodland, open heath and grassland and enclosed farmland.

View from Boat How looking east

Your viewing position: Looking north-east across the lake

Ennerdale Water: Ennerdale is a large glaciated valley and from where you are sitting you have a good view of two roche moutonnées that leave no doubt that the ice moved from right to left – one is the huge bulk of Bowness Knott and the other a much smaller feature at Bowness Point on the shore below. The curious shape of Ennerdale Water itself, with the upper two-thirds looking like a typical ribbon lake in a glaciated valley contrasting with the broad rounded western end, is largely the result of underlying geology. The upper part of the lake and much of the Ennerdale valley have been excavated by a glacier in Ennerdale granite, but the lower end of the lake lies in less resistant Skiddaw Group rocks where glacial erosion could take place across a broader front. However, Angler's Crag and the west side of Bowness Knott, despite being composed of Skiddaw Group rocks, have resisted erosion because they have been metamorphosed and hardened through contact with the igneous granite intrusion. Bowness Knott is designated a Site of Special Scientific Interest on account of this interesting geology. After the Ice Age, thick deposits of glacial till impounded the water to create the lake and influenced the shape of the shoreline and the location of the River Ehen outflow.

Water extraction: Ennerdale Water is owned by United Utilities and water is extracted to help meet the water requirements of West Cumbria. Water was first extracted in the mid-19th century and in 1902 a small weir was constructed to maintain the lake's water level, the natural outflow being over relatively soft glacial till. In 1978 there was a proposal to raise the water level to provide additional water supplies for the nuclear and other industries in West Cumbria. This proposal was met with strenuous opposition from a wide range of amenity and recreational groups because of the environmental impact of the planned reconfiguration of the northern shore and the visual impact of changing water levels at the strandline. The proposal failed to get permission to proceed.

In 2015 the Environment Agency served notice on United Utilities to cease water extraction from Ennerdale Water by 2022 because the amount of water being taken was judged to be adversely affecting water flow and habitats downstream in the River Ehen. The river is designated an internationally important Special Area of Conservation, primarily because of the presence of England's largest breeding population of freshwater mussels (a statutorily protected species) and the habitats that support them. At considerable cost and amidst some public controversy, United Utilities is building a new water pipeline from Thirlmere to West Cumbria to make up for the loss of supplies from Ennerdale Water.

The Side: Where you are sitting is within a very large, roughly rectangular walled enclosure known as The Side. It was a deer park in the 17th century (and probably for some time before) where, it was recorded in 1675, there could be found "Hartts and Staggs as great as in any part of England". In fact the deer park was in its final years; it was but a small remnant of the huge Copeland Forest hunting grounds that had gradually been given up in the face of an expanding population and demands for land for sheep grazing and agriculture. This tract of land lies within the Pillar and Ennerdale Site of Special Scientific Interest, which according to the citation is "important in exhibiting one of the best known examples of altitudinal succession in England", which essentially means changes in natural ecosystems in response to elevation. Directly below you and to your right, at around 150 metres, is Side Wood which is native birch-oak woodland supporting some relatively rare lichen communities; above, where Boat How is located, is sub-montane vegetation of extensive heathland dominated by heather with small areas of acid grassland; above this the heaths become montane in character dominated by bilberry and grasses and, higher again, along the summit ridge the thin soils support a mixture of bilberry and moss-dominated heaths.

Settlement history: Ennerdale is notable for the low intensity of past human occupation and farming, which has left a rich legacy of settlement remains, many protected as scheduled ancient monuments. Looking to your right up-valley, just beyond where the River Liza comes out of the present forested areas, there is a field clearance cairnfield that is possible evidence of Bronze Age occupation, and in the same vicinity there are remains of an enclosed Romano-British settlement marked as a "homestead" on the Ordnance Survey map. Nearby, the names "High" and "Low Gillerthwaite" suggest a Norse presence ("thwaite" meaning a clearing), but there is no visible evidence. In the medieval period there was a sizeable agricultural community just beyond High Gillerthwaite. The settled area lay within the present wooded areas just up and left near the far end of the open green valley floor. It is thought to have been a community living in twelve rectangular longhouses with cultivation terraces and pounds for stock that extended up the north side of the valley. At Gillerthwaite and just beyond the far end of Ennerdale Water where Woundell Beck joins the River Liza, there is documentary evidence of vaccaries in the 14th century, probably reliant on high pastures, such as those in Great Cove. During the medieval period there was very little enclosure, with the notable exception of The Side. In the post-medieval period there was an adjustment of agricultural settlement with the establishment of the Gillerthwaite farms and the building of the valley-bottom enclosures, but avoiding the earlier almost adjacent settlements. The enclosure of Ennerdale in the 1870s, with long straight boundaries that determined later forestry layouts, was one of the last large-scale Parliamentary Enclosures in England.

Remains of a longhouse at Smithy Beck settlement, probably late medieval

Iron and a railway: Opposite you, on the far side of Ennerdale Water, is Smithy Beck. As its name suggests, this was the focal point for a late-medieval iron-working landscape of mines, bloomeries and longhouses that were probably the homes of iron workers. The iron ore was mined at Clewes Gill, a tributary of Smithy Beck, and taken down to the shore of Ennerdale Water where there was an iron smelting bloomery and charcoal burning sites. The site of the assumed iron workers' settlement is shown on the Ordnance Survey map, close to the confluence of Clewes Gill and Smithy Beck. All these remains are scheduled ancient monuments.

Interest in Ennerdale's iron-ore reserves grew further during 19th-century industrialisation, with a number of new small and exploratory mines and trials in the area adjacent to Boat How and westwards over Crag Fell. In order to exploit the valley's iron-ore resources fully, a mineral railway line to the head of Ennerdale Water was proposed in 1883. This was strenuously and successfully opposed by the newly created Lake District Defence Society and others. This was a welcome victory for conservationists (still licking their wounds after losing the battle over Thirlmere a few years earlier) that helped build a wider acceptance of the principle that development proposals in the Lake District needed to take account of landscape and other environmental impacts.

March of the conifers: The views in front of you are dominated by forests and for much of the last 100 years Ennerdale and commercial forestry have been synonymous. Following the end of the First World War, the Forestry Commission was set up with the remit to rebuild the country's timber resources and provide rural employment. In 1925 the Commission acquired 5,000 acres of land in Ennerdale and within a few years the valley was blanketed by regimented rows of larch and spruce arranged in unnatural straight-edged blocks. Agriculture was excluded and opportunities for wildlife and recreation diminished. The product of the single-minded forestry practices of the time was plain for all to see and disapproval was widespread. The Ennerdale conifer plantations provided the alert, showing what could happen to all the central Lake District fells, and this ultimately led to an historic agreement in 1936 between the Commission and the Council for the Preservation of Rural England. This hard-won agreement is explained in the pages of this book on Dod Knott, Eskdale and Castle How, Upper Duddon. Forestry Commission practices today are a world away from those of the 1930s and some of the most innovative are now in play under the auspices of the Wild Ennerdale project.

Wild Ennerdale: The Wild Ennerdale Project was begun in 2003 as a multi-decade project to allow the valley to evolve as a wild valley through increasing reliance on natural processes over time. On this scale the project is pioneering and outside the mainstream of rural land-management practice in England. The project is run as a partnership of the Forestry Commission, National Trust and United Utilities (the three landowners) and Natural England. It has as its overall vision "To allow the evolution of Ennerdale as a wild valley for the benefit of people, relying more on natural processes to shape its landscape and ecology". The project is not intended to recreate a past landscape and ecology but to allow natural processes to determine the valley's future character. The landscape is not being "managed" in the usual sense, although there is some intervention where necessary to encourage natural processes, such as removal of structures affecting the flow of the River Liza, and to support struggling native species such as the Arctic char, the red squirrel, the juniper and the Marsh Fritillary butterfly.

Volunteers restoring water levels in the previously drained Gillerthwaite Mire

Over recent years, the project has worked on reducing the dominance of Sitka spruce
through the clearance of young planted and regenerating spruce, prioritising felling
of visually intrusive areas, planting native broadleaves and allowing natural woodland
encroachment outside currently forested areas. Three herds of Galloway cattle, managed
by tenant farmers, have been introduced and their extensive grazing has greatly influenced
habitat change. Within the valley cattle are now grazing more than half of the total wooded
area, providing a glimpse of how pastoral farming, woodland management and habitat
renewal can be integrated.

Under the Wild Ennerdale vision, the view across Ennerdale Water from where you are sitting
will gradually change. Much of the woodland from below Bowness Knott rightwards will be
managed through low-impact silvicultural systems that aim to produce more diverse forest
structures with less spruce and more Scots pine, birch and other species, and with more
open woodland along watercourses and around historic sites. On Bowness Knott the exposed
clearfell will be replaced with native woodland and scrub (some planted and some natural
regeneration) and some planted juniper. To your right on the far shore at the top of the lake
there is an area of native deciduous ancient woodland where operations will be restricted to
those that sustain and enhance its current characteristics. Further right up-valley beside the
River Liza is what has been termed a Minimum Intervention Riparian Corridor where there
will be no felling or planting of trees to allow natural river processes (including changes of
course) and natural tree and scrub regeneration. Finally, along the complete length of shoreline
opposite you, there is potential for new wetland habitat to evolve as water levels fall following
the end of the era in which the lake functioned as a reservoir.

Brown Howe, Mardale

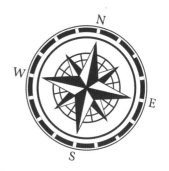

A good viewpoint looking over the Haweswater reservoir and surrounded by land managed by the Royal Society for the Protection of Birds

Location:	NY483124; N54°30.233 W002°48.013; just above Brownhowe Crag, a little below the uncairned summit of Brown Howe
Height:	450 metres
Geology:	Borrowdale Volcanic Group
Ownership:	United Utilities
Access:	Open access land; traverse across rough ground from the Old Corpse Road
Website panorama:	From south through west to north

Brown Howe from the west

Location of Brown Howe

♀ The View

Centre stage is the Haweswater reservoir with its enclosing fells well described by the Cumbrian poet Norman Nicholson as "standing waist deep in water". It lies within a dramatic enclosed valley head tucked under High Street that when covered in deep snow looks as alpine as anywhere in the District. The long, curving body of water fully occupies the width of the valley, displacing all signs of past settlement and farming; even the fellsides are now only sparsely grazed by sheep. It all adds up to a valley with a strong atmosphere of isolation and emptiness. Within this setting the RSPB, the main common rights holder on Mardale Common, and landowner United Utilities hope to demonstrate ways of achieving a more sustainable approach to pastoral farming in the Lake District uplands.

View from Brown Howe looking south-west

Your viewing position: Looking west across the reservoir

Corries: The spectacular, glaciated scenery of the valley head in front of you and to your left is in a classic corrie-forming location facing east in the lee of the High Street range. The fine corrie containing Blea Water is a textbook example of where the rotational slipping of ice over numerous glacial periods has gouged out a rock basin now containing an astonishing water depth of 63 metres (almost the same as the deepest point in Windermere). This scenery on the east side of the High Street ridge contrasts markedly with the smooth profiles seen on the ridge itself and its west-facing flanks, replicating the differences seen on the western and eastern flanks of Helvellyn described and explained earlier in the section on the physical landscape. The corries and valleys you are looking into (Small Water, Blea Water, Riggindale and opposite you in Whelter Bottom) are "eating" into an earlier landscape, giving us a fascinating insight into how this and other parts of the Lake District might have looked before the ice got to work.

Valley settlement: Archaeological evidence scattered around much of your view suggests fairly continuous occupation since prehistoric times. On Castle Crag opposite you there are remains of a small hillfort and massive stone bank that are probably of immediate post-Roman date. Below and just right of the fort is a prehistoric cairnfield and on Four Stones Hill, far to your right just within view, there are a Bronze Age enclosure, clearance cairns, burial cairns and standing stones. Across the reservoir in Riggindale there are remains of an enclosure, hut circles and a platform that are possibly of prehistoric date. Along the skyline above the head of the valley a Roman road runs on High Street northwards towards the outskirts of Penrith, and during the Roman period there were Romano–British settlements lower down Mardale at Measland Beck (now underwater), near the north end of the dam and also south of the Haweswater Hotel.

During the medieval period lands were granted to the monks of Shap Abbey who developed sheep farming and wool production in the Lowther valley and made use of the higher pastures in Mardale. The valley was settled and farmed along the northern shores of the original Hawes Water and in the upper part of the valley. There is evidence of medieval buildings at Whelter Beck and elsewhere, and records suggest a common field, Mardale Field, to your left in the upper part of the valley on both sides of Mardale Beck. Throughout the medieval period and until the route over Shap was favoured and improved in 1753, the valley would have been busy with animals and goods passing through on the important packhorse and drove routes between Kendal and Penrith that went over the Gatesgarth and Nan Bield passes at the valley head. You probably approached Brown Howe from the Old Corpse Road. This was the route for carrying bodies out of the valley for burial in Shap during the medieval and post-medieval period; Mardale did not have a consecrated burial ground until 1728. You may also have noticed a number of ruined buildings close to the Old Corpse Road, just above the steep descent into the valley. The ages and purposes of these buildings are uncertain; possible uses were as bothies and peat huts.

Early tourists: The valley's contribution to the wider area's prosperous wool trade continued into the 18th century, but when other parts of the District were beginning to realise the economic opportunities provided by the early tourists, Mardale never really got in on the act. The valley did not excite the attention of the seekers of the Picturesque and was not a place chosen for villa building in the 18th and 19th centuries. Doubtless Thomas West's description of Mardale Head in his 1778 *Guide to the Lakes* as "all is hopeless waste and desolation strewed with the precipitated ruins of mouldering mountains" did nothing to encourage the visitors! However, the valley's attractions were appreciated by a few, and in 1811 the Leeds industrialist John Marshall bought Low Whelter, a farm now under water near Whelter Beck, opposite you. This was his first purchase in the Lake District, bought for its scenic setting and the opportunity to manage some attractive woodland (more on John Marshall can be found in the pages on Lambing Knott and Brown Knotts). His mentor on

Coloured engraving of Mardale Green around 1830

woodland design, William Wordsworth, equally appreciated Mardale because "it remains undefiled by the intrusion of bad taste". No prizes for guessing what he would have said about the taste of those who built the reservoir dam.

Old Mardale: Before the valley was flooded in the 1930s there was a community and a different landscape. The point where Whelter Beck reached the valley floor was the south end of the old Hawes Water and it extended northwards to roughly where the dam is built. It was almost divided into two by the Measand Beck delta, with the two parts named High Water and Low Water. The road up the valley ran along the north shore, so a traveller coming into your view on your far right passed through the small hamlet of Measand on the delta and then continued past the occasional dwelling until under Whelter Knotts where the school was sited. Then the former farms of Low and High Whelter were passed, the Parsonage under Castle Crag and a little further on the farm of Flakehow. The road then took the traveller over Riggindale Bridge and into the main village of Mardale Green, passing the cluster of buildings at Chapel Hill and Holy Trinity Church between The Rigg and Wood Howe. Then on to Chapel Bridge to cross Mardale Beck at the north-east end of The Rigg, past the farms of Goosmire and Grove Brae and other buildings to arrive at the Dun Bull Inn east of The Rigg. Here the road ended but the historic packhorse tracks continued up over the Gatesgarth and Nan Bield passes.

Crowds outside Mardale church for the farewell service on 18th August 1935

Haweswater reservoir: The reservoir swept all this away. Prior to the flooding of the valley in 1935, when there were 40 residents living in the valley, the four farms, nine inhabited dwellings and the community buildings were demolished and 104 coffins were removed from the graveyard for reburial at Shap. The Manchester Corporation Act 1919 gave the council the power to acquire the land within the Mardale watershed, build the dam and flood the valley by raising the water level by 29 metres. There was a fair amount of protest when Manchester's plans were first announced about the loss of public amenity and the two valley communities, but the protests were not on the same scale as those against Thirlmere, presumably because the valley's scenic qualities were not so widely appreciated. In the event, the protests did lead to obligations within the Act to construct a public footpath along the north shore and to allow public access on all common and unenclosed land within the watershed – thanks to a remarkable piece of lobbying by the Commons, Open Spaces and Footpaths Society decades before "right to roam" legislation came along. The Haweswater Hotel was also built as a replacement for the Dun Bull Inn, a long-appreciated local hostelry and venue for the annual Mardale Shepherds' Meet. Like the Thirlmere scheme, Manchester Corporation's plans included extensive forestry planting around the reservoir, but in this case for some reason the planting was never implemented.

Water is taken from the reservoir at the draw-off tower down to your right just out of sight (but marked on the Ordnance Survey map) and first started to flow in 1941 down the initial section of aqueduct that connected with the Thirlmere aqueduct near Kendal. The aqueduct is in a tunnel almost under where you are sitting. Down to your right you can see one of the original aqueduct survey pillars and way up to your left there is another on the skyline col between Artlecrag Pike and Selside. The Haweswater aqueduct finally completed its independent route to Manchester in 1955. After further extensions to the aqueduct network in the 1970s, it now carries to Manchester double the amount of water carried by the Thirlmere aqueduct and provides 25 per cent of North-West England's water requirements.

RSPB upland farming trial: You are sitting in the middle of a landscape that is subject to an ambitious, and in some quarters controversial, upland farming trial. All the land in the Haweswater catchment is owned by United Utilities, which agreed in 2012 to the RSPB taking over the tenancies for Naddle Farm in Mardale and Swindale Farm in adjacent Swindale. Naddle Farm's land is in two parcels: all the enclosed land around Naddle Beck some distance to your right and all the enclosed land at Mardale Head and lower Riggindale to your left. Additionally, the farm has extensive grazing rights on all the unenclosed land within your view and beyond (Mardale and Bampton commons). The RSPB's overall aim is to evolve a more sustainable approach to farming in the uplands through trials of different approaches to upland management, with the Naddle and Swindale farms operating as a single business unit. The key objectives are to run a hill farm and at the same time restore biodiversity, improve drinking-water quality and increase carbon storage. This involves environmental projects to stabilise soils, plant native trees and plants, and restore peat and other habitats, but at the core of the experiment is trialling different sheep stocking levels and measuring the impacts in relation to the project's four objectives. The RSPB has undertaken to produce periodic economic reports on the farm's business performance and the report for the three years 2013 to 2016 presents a stark summary of the challenge: income from sheep sales is less than a third of sheep production costs and the farming business only shows a profit when its substantial European grants are added into the balance sheet. But what the RSPB-commissioned assessment of the farm business points out is that, subject to a break-even threshold, reducing stocking rates on this farm would improve business viability because costs would fall at a greater rate than income.

It has to be said, however, that this RSPB work, while supported by many as the way ahead, is controversial in a number of quarters. Some fear that the RSPB is creating in Mardale and adjacent Swindale (see pages on Thorny Knott) a testing ground for new landscapes that could in due course be rolled out across the Lake District with potential adverse consequences for upland sheep farming and its traditions, the livelihoods of farming families and the appearance of the World Heritage Site's celebrated pastoral landscape. The counter view is that taking steps to enhance biodiversity and secure other environmental improvements has now become urgent and can be assisted through less-intensive pastoral farming models.

The management plan for the World Heritage Site prepared by the Lake District National Park Partnership expresses the issue as "reconciling tensions between managing the cultural landscape and enhancing the natural environment". The lessons learnt from the management of the land all around you can be expected to contribute in some way to finding out how this "reconciling" might be achieved.

Brown Knotts, Derwent Water

A prominent position well above the eastern shore of Derwent Water providing a spectacular vantage point over the lake and lower reaches of Borrowdale

Location:	NY274199; N54°34.231 W003°07.403
Height:	350 metres
Geology:	Borrowdale Volcanic Group
Ownership:	National Trust
Access:	Open access land; accessible by descent from Bleaberry Fell or from the rising footpath from Ashness Bridge
Website panorama:	From north through west to south

Brown Knotts, just below centre skyline, from the west

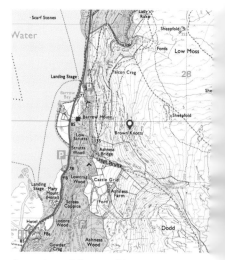

Location of Brown Knotts

⚲ The View

The broad expanse of Derwent Water and its islands dominate the view with the town of Keswick on its north shore, the gentle fell of Catbells on the west and the popularly dubbed "Jaws of Borrowdale" to the south, all enclosed within an amphitheatre of more distant high fells. Here the ravages of past industry have been healed so that today water, woods, fields, villas, fells and sky combine to create a spectacularly beautiful and tranquil scene. It is a landscape that inspired poets and artists, fostered new ways of seeing and appreciating the natural environment and attracted early tourists by the trainload.

View from Brown Knotts looking south-west

Your viewing position: Looking west across the lake towards Catbells

"Overhanging" rocks: Brown Knotts is part of one of the tiers of spectacular volcanic crags, from Walla and Falcon Crags on your right to King's How on your left, at the geological boundary with the sedimentary Skiddaw Group rocks of the Derwent Fells in front of you. In the late 18th century, a visit on horseback along a rough track under these crags, leading into what came to be known as the "Jaws of Borrowdale", would have been viewed as a bold undertaking. It was the descriptions by early visitors and writers of these fearsome crags and the adjacent scenes of natural beauty that did so much to develop a national fascination with Lakeland scenery. Here could be experienced scenes of both the Beautiful and the Sublime. Thomas West's guide, published in 1778, conveys the mixture of scenes and emotions:

The whole of the western extremity [of Derwent Water] is beautiful beyond what words can express, and the north end exhibits what is most gentle and pleasing in landscape. The southern extremity is a violent contrast to all this. Falcon crag, an immense rock, overhangs your head, and upwards, a forest of broken pointed rocks, in a semicircular sweep, towering inward, form the most horrid amphitheatre that ever eye beheld in the wild forms of convulsed nature.

How could anyone with a modicum of adventurous and curious spirit resist visiting such a place? Quite possibly it was the writings about this small tract of rocks that did most to set the Lake District tourist industry on its way.

An engraving of 1827 looking south towards the Jaws of Borrowdale conveys "the most horrid amphitheatre"

KESWICK-LAKE.

The Lake: Derwent Water is one of the District's environmental gems and an integral part of the River Derwent Special Area of Conservation. The catchment of the River Derwent is considered to be the finest example of a large oligotrophic (low nutrient) river system in England. Derwent Water itself is noted for displaying rare aquatic flora, shorelines with undisturbed habitat transitions from open water to reed swamp and carr woodland, and for the nationally rare vendace, a relict fish species isolated in the lake after the last Ice Age. Although Borrowdale is one of the linear, radiating valleys of the Lake District, Derwent

Water is not one of the typical narrow ribbon lakes found elsewhere. The lake basin was gouged out by a north-flowing glacier, but rather than a narrow trench we have a lake that is broad and oval in shape, some 1.7 km wide and 4.5 km long. After the glacier squeezed its way through the hard volcanic rocks of the "Jaws of Borrowdale" (between Castle Crag and King's How), it then eroded the relatively soft rocks of the Skiddaw Group on a much broader front creating a shallow rock basin and a lake with an average depth of only 5.5 metres. The glacier deposited glacial till that was shaped into drumlins forming the islands to your right – Derwent, Lord's, St Herbert's and Rampsholme – and the low hills of Stable Hills, Cockshot Wood and Crow Park. Glacial till also forms the sill at the northern outflow from the lake, a feature which is slowly eroding and giving cause for concern about the water level, not least for recreational boating.

The islands: St Herbert's Island is named after the saint, a disciple of St Cuthbert of Lindisfarne, who used the island as a hermitage. Lord's Island, once owned by the Earls of Derwentwater, had a large house built on it around 1460. The house and the drawbridge that connected the island to the shore have now gone. Derwent Isle was occupied for a period in Elizabethan times by German miners working in the area, apparently as a refuge from some of the local populace who were not so enamoured with the miners' interest in the local young women. In 1778 Joseph Pocklington built a mansion and various follies on Derwent Isle and it is now the only island still occupied. Rampsholme Island may have been the site of a medieval iron bloomery. All four islands are now owned by the National Trust.

Elizabethan mining: Opposite you, on the western shore and over Hause Gate into the Newlands valley, was the setting for the extraordinary expansion of minerals exploitation in the Lake District during Elizabethan times. In 1564 the Company of Mines Royal was set up to exploit metal ores in various parts of the country and experienced German miners were brought in to provide the necessary expertise. Activity was immediately underway in the Newlands valley and on the western shore of Derwent Water: notably copper mining in the vicinity of Swanesty How, west of Grange, and lead mining on the eastern flank of Skelgill Bank on Catbells. Copperheap Bay on the north-west shore is where large volumes of copper ore mined in Newlands were ferried across Derwent Water en route to a smelter at Brigham on the east side of Keswick. The Brigham smelter was a huge operation, the largest in the country, with six furnaces requiring vast quantities of charcoal that was produced by decimating not only local Borrowdale forests, but also those much further afield. Smelting stopped around 1650. However, mining continued for the next 200 years and by the Victorian era was on a very large scale, contributing to the scenes visited by the earliest tourists despite rarely being commented on. One of the largest enterprises was Brandelhow mine with its main buildings located on Brandelhow Bay almost directly opposite you. Here, good-quality lead ore was mined although operations were regularly hampered by flooding that affected the mine's financial viability. With a very large waterwheel, steam pumps, crushing and dressing sheds and other buildings, the site must have been quite a scene. In 1891 the site was abandoned.

Woodlands lost: Within much of your view, and to your left in particular, there are extensive areas of ecologically rich woodland, much of it designated Sites of Special Scientific Interest. Their abundance raises the question about why such fine woodlands are here but largely absent in the adjacent valleys of Newlands, Watendlath and Langstrath. The woodlands around Derwent Water and in the adjacent valleys had been exploited from the earliest periods of human settlement, and by the mid 18th century, exploitation for charcoal to smelt iron and other ores, like lead and copper, had resulted in substantial deforestation. However, an event in 1749 on the shore of Derwent Water brought about a change in direction. The Commissioners of the Royal Greenwich Hospital in London managed all the land around the north-east shores of Derwent Water, to your right, as part of their northern estates that generated income from land rents, timber and mining to support the country's Royal Navy pensioners in need. Part of their Keswick estate at Crow Park, between the town and the lake, was covered with mature oaks that the Hospital decided to fell and sell for timber in order to meet some pressing financial commitments. The felling started in 1749, which led immediately to considerable public criticism of the Hospital's estate-management policies and fuelled a wider debate about landscape aesthetics and the impact of private decisions on the interests of the wider public. In retrospect, it can be seen that this controversy triggered an environmental awakening in the Lake District which later led to the rise of the organised conservation movement.

Woodlands regained: While the chastened Hospital continued to manage the eastern shores with greater emphasis on planting than felling, on the western shores of Derwent Water Lord William Gordon (brother of Lord George Gordon, who was imprisoned for instigating the Gordon Riots in London) set about the creation of Picturesque scenes for public enjoyment. From 1781 to 1787 he purchased much of the western shore from Fawe Park to south of Brandelhow. He built a pavilion at Derwent Bay (now replaced by Derwent Bay House) as part of a designed Picturesque park that was for public appreciation and enjoyment rather than his own personal use. Here was a display of Lord William Gordon's good taste and public spiritedness that has left an important woodland legacy. Back on the east shore, in 1832 the Greenwich Hospital finally decided to sell up and all their land on Derwent Water was sold to John Marshall, a wealthy Leeds industrialist. He was a man of the moment who blended a love and appreciation of natural scenery with equal commitment to industrial progress, improvement in society and financial returns. All the land in your view on the east side of Derwent Water from Keswick to Barrow Bay was in his family ownership and, with advice from William Wordsworth, extensive woods were planted in a manner that sought to enhance the naturalness of the scenery.

In due course much of the environs of Derwent Water, except on the north-west shore, came into National Trust ownership. Brandelhow, acquired in 1902, was the first of the Trust's land purchases in the Lake District. The woodlands around Derwent Water and in Borrowdale have been carefully nurtured and now thrive in scenic splendour; adjacent valleys do not have the same history of benefactors' care and attention.

Official opening of the National Trust's newly acquired land at Brandlehow in 1902

Keswick and tourism: Far to your right at the north end of the lake is the town of Keswick, which can probably claim the distinction of having become the first tourist centre in the Lake District. On 16th September 1783 its tourist industry got underway with a bang – literally. A large regatta and a mock military invasion of Derwent Isle was organised to provide additional entertainment for the well-to-do visitors, supposedly here to educate themselves in the aesthetic qualities of the Derwent Water and Borrowdale scenery. This entertainment was a joint enterprise by two local notables: Joseph Pocklington and Peter Crosthwaite. Pocklington was the son of a wealthy Nottinghamshire banker and had built a mansion on Derwent Isle and Crosthwaite, an early convert to the lucrative potential of tourism, acted as a guide, produced local maps and established a museum in Keswick. By 1802 the poet Samuel Taylor Coleridge, who was living in Greta Hall in Keswick, observed the town was "swamped with tourists". Coleridge and his fellow Lake Poets, Robert Southey (who moved into Greta Hall in 1803) and William Wordsworth (a regular visitor), wrote and published poetry extolling the beauties of the area and did much to fuel their readers' urge to visit. But it was in 1864, when the Cockermouth, Keswick and Penrith Railway opened, making the area more readily accessible to a broader section of society, that Keswick's tourist industry really took off. The demand for accommodation for new residents and visitors grew substantially during the latter part of the 19th century and with the construction of new hotels, guest houses and residential properties the town developed rapidly with a distinctive Victorian character. Today the town is the focus for the tourist industry in the north Lakes.

Castle How, Duddon

A prominent rocky outcrop overlooking a tranquil and timeless Duddon valley scene

Location:	SD189921; N54°19.087 W003°14.852
Height:	144 metres
Geology:	Borrowdale Volcanic Group
Ownership:	Private
Access:	Immediately adjacent to a public bridleway
Website panorama:	Two panoramas giving a 360° view clockwise from north

Castle How from Ulpha Bridge

Location of Castle How

The View

A patchwork of ancient woodland, meadows, rough pastures, rocks and scattered farmsteads intertwined with the meandering river create a delightful valley scene. Although glaciated, this valley is more intimate than the deep and spacious glacial troughs elsewhere; it is the woodlands rather than precipitous rocky fellsides that create a sense of enclosure. This is a landscape, more densely populated in the past than today, shaped by communities engaged primarily in pastoral farming and a variety of woodland industries. Never really on the itinerary of the seekers of the Picturesque, the valley was nevertheless much loved and written about by William Wordsworth.

View from Castle How looking north-east

Your viewing position: Looking north-east up the valley

A lake-less valley: Unlike nearly all the valleys radiating from the central fells, the Duddon valley does not contain a large lake; nor are there theories about old lakes that have been filled in. Looking at the pattern of radiating ribbon lakes in the Lake District, there is a marked gap in the sector from Coniston Water westwards round to Wast Water. This cannot be readily explained by underlying geology because, across the Lake District as a whole, there appears to be no particular correlation between the locations of the large lakes and the rock type within which their basins have been gouged out. During the Ice Age, it is thought that the ice flows from the high central fells down towards the south-west were more active than elsewhere in the District because of higher precipitation and unimpeded convergence with the south-flowing ice from Scotland in the Irish Sea basin. Why this ice movement did not erode a deep rock basin in the Duddon valley (or in neighbouring Eskdale) has not been satisfactorily explained, but may have had something to do with the nature of the pre-glacial topography. This may well have consisted of shallow valleys with low, broad interfluves over which the ice could move easily rather than concentrating all its erosive power within a narrow valley confine. Whatever the reasons, the consequence for the Duddon valley was that the absence of a lake meant there was no natural focus to attract the searchers of Picturesque beauty or encourage wealthy outsiders to build their villas with designed gardens. The valley has remained a relative backwater, untroubled by large numbers of visitors. The quiet serenity of the valley was, however, much appreciated by William Wordsworth, and his celebrated Duddon Sonnets published in 1820 convey its distinctive character and moods from source to sea.

Early human occupation: Although there is no direct evidence, it is highly likely that humans were in the Duddon valley in the Neolithic period; the impressive Swinside stone circle, dated to the latter part of this period, lies only four kilometres to the south-west. Then with a warming climate at the start of the Bronze Age, 4,000 years ago, humans started to occupy lower fells in the District just above the wet and still densely wooded valley floors. A typical location is the land around you. Turning leftwards from where you are sitting, the steep valley side covered by Rainsbarrow Wood lies back to form an extensive tract of less-steep land at an height of 150 to 250 metres between Holehouse Gill and Logan Beck to the south. Below The Pike, the land at Pike Side Farm coming down towards you was surveyed as part of a Duddon valley community archaeology project called Ring Cairns to Reservoirs. The investigations here found evidence suggesting intermittent occupation from prehistoric times through to the medieval period, notably clearance cairns, a hut circle, field patterns, enclosures and a group of three possible medieval longhouses constituting a small settlement.

Medieval Ulpha Park: During the medieval period there would have been farmsteads on the land to your left, but a dominant feature in the landscape would have been an enclosed deer park that extended from Rainsbarrow Wood down to Logan Beck. This land is marked as Ulpha Park on the Ordnance Survey map. "Park" is a term of medieval French origin – "parc" meaning enclosure or fenced-off area. The creation of these parks began in the 13th century when it made more economic sense for parts of the forest lands to be made available for grazing sheep and cattle, and to restrict deer to these more limited enclosed areas. In the 14th century the Hudleston family of Millom Castle established Ulpha Park within the Forest of Millom, which covered much of the Duddon valley. The deer park served the practical purposes of providing meat and sport for the lord's family and guests, while also serving as a symbol of wealth and status.

Frith Hall: Behind you, within the confines of Ulpha Park, are the gaunt ruins of Frith Hall. It was probably built in the 16th century as a hunting lodge for the Hudleston family and may have replaced the old fortified house, Ulpha Old Hall, now named Old Hall Farm, a short distance to your left. Subsequently the Hudlestons found themselves on the wrong side in the English Civil War and departed from Frith Hall, which later became an inn of ill

Remains of Frith Hall

repute as a refuge for eloping couples, runaway marriages and various dubious activities. By the early 19th century, in Wordsworth's day, it had become a farm, later to be abandoned as a consequence of changes in the farming economy.

Enclosure: In the post-medieval period the more usual picture in the upland valleys of extensive, flat inbye land on the valley floor with walled intakes rising up the lower slopes on both sides of the valley is not the case here. The difference in appearance between the east and west sides of the Duddon at this point is marked. Across the river to the east, the unenclosed rough fellside still retains its common-land status and comes right down to the roadside on the valley floor. On the west side the amenable topography, which had encouraged settlement in the Bronze Age, was settled during the medieval period in the vicinity of Holehouse Gill, Tongue Beck, and the Blea Beck and Logan Beck tributaries, with an area of medieval inbye land considerably greater in extent than that on the nearby valley floor. During the 16th and 17th centuries the former open fields, subdivided within families to the point of poverty, were reorganised through the manor courts and enclosed as strips, as can still be discerned from the pattern of walls in the area between Whinfield Ground and Logan Beck Bridge. Subsequently farms amalgamated and some farmsteads were abandoned. Surveys suggest that the only old intakes are small areas upslope from Hole House and Pike Side farms and that other enclosed land, including close to Castle How and down to the river, was enclosed relatively recently, during the last two hundred years.

Worthy of note is the alignment of the present minor road that runs from Ulpha westwards to Old Hall Farm, then south and back east to the valley floor. This off-shoot from the main valley road, serving numerous farms and dwellings, is unusual and reflects the favoured use of this slightly elevated land for settlement from very earliest times to the present day.

Ancient woodland: From where you are sitting the views along the western side of the valley are dominated by woodland. Much of this woodland is part of the extensive Duddon valley Woodlands Site of Special Scientific Interest (SSSI) that stretches from Duddon Bridge all the way up the valley to Seathwaite. This is one of the largest series of woodlands in the Lake District. Most is categorised as ancient semi-natural woodland displaying a particularly wide and distinctive range of habitats and species. The most widespread trees are sessile oak, birch and some sycamore. Looking north, all the woods you can see on the west bank of the river, except the southern part of Rainsbarrow Wood, are within the SSSI; looking south, so too are much of Forge Wood, Middle Park and Low Park. Other woods in your view, outside the SSSI, at Birks Wood on the east side of the Duddon, the southern part of Rainsbarrow Wood and part of Middle Park are what are termed plantations on ancient woodland sites, currently dominated by non-native species but subject to management regimes intended gradually to restore them to native woodland.

All the woods within the SSSI are periodically assessed by Natural England for their condition. At the most recent surveys (2009 and 2013) only Stonegarth Wood, which runs up beside Holehouse Gill, was deemed to be in "favourable" condition. All others were judged to be in an "unfavourable" condition, but with Rainsbarrow Wood, Pike Wood and Forge Wood judged to be "recovering" in response to active management remedies. Each wood has

Chimney (ivy covered) at the site of the former bobbin mill on Holehouse Gill

its own particular management issues but a common problem is grazing by deer, which is inhibiting regeneration.

Woodland industries: The character of the woodlands you see and, indeed, their continued existence through to today is substantially because of their role in supporting past industries and the efforts that went into managing and protecting them. The first clues to these past activities are to be found marked on the Ordnance Survey map – immediately to the south of Castle How is Forge Wood, to the east across the river is Cinder Hill with its "ironworks" and to the north of Castle How a "mill" on Holehouse Gill. From medieval times these woodlands were the source of coppice wood for making charcoal at round pitsteads; remnant examples can be found in Rainsbarrow Wood. Supplies of iron ore were brought to the charcoal (not the other way round) and smelted at various small iron bloomeries, such as on the east side of the Duddon at Cinderhill Beck and a little further south at Stonescar, and then by the 17th century at a larger forge at Forge Wood operated by the Hudleston family. These iron-production sites were eclipsed after 1736 by the massive charcoal-fuelled blast furnace built a few miles downstream at Duddon Bridge. The woodlands also produced a range of wood products including bobbins for the Lancashire cotton industries. A chimney still stands to indicate the site of the mill on Holehouse Gill – a former water-driven bobbin mill that produced bobbins for Lancashire Thread mills until 1910.

The Almshouses: Prominent on the valley floor just north of Ulpha is a semi-circle of single-storey dwellings. These are almshouses, possibly the only ones in the National Park, built in 1914 by a trust set up and bequeathed by John Gunson of Oak Bank, Ulpha, in memory of his parents. The trust provides low-cost housing for those "in need and of good character" who were born in Ulpha or in the vicinity. Although established a hundred years ago, the work of this trust continues to have a real relevance today in the Duddon valley and throughout the Lake District as many local people are priced out of the local housing market. The National Park Partnership's current plan for the Lake District contains a suite of policies to support "vibrant communities", including the provision of affordable housing for local people. A local housing initiative like this helps sustain the vitality of community life, but it is an uphill task – Ulpha Primary School, amongst the buildings just east of the bridge, had to close in 1999 because pupil numbers were too low.

Castle How, Upper Duddon

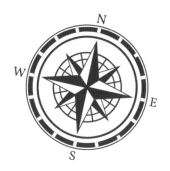

A huge outcrop of rock on the valley floor with a summit
giving views of the upper Duddon valley and progress on the
transformation of Hardknott Forest

Location:	NY237004; N54°23.660 W3°10.572
Height:	273 metres
Geology:	Borrowdale Volcanic Group
Ownership:	Forestry Commission
Access:	Open access land; the summit can be gained either from the west to gain the neck between the two Castle How summits or from the north-east to the col between Castle How and Saddlebacked How; easy scrambling to the summit
Website panorama:	From south through west to north-east

Castle How from the south

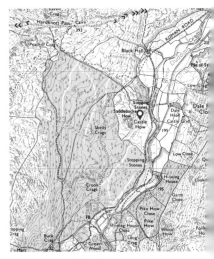

Location of Castle How

♀ The View

This is a glaciated landscape of wild, craggy fell country many miles from the nearest
settlement. But thanks to the precedent set by Roman road builders, it is also very accessible

with a regular stream of cars enjoying the challenges of the roads over the Wrynose and Hardknott passes. It is in winter after snowfall when the passes are closed that the true remoteness of this place becomes sharply apparent. The valley, with its occasional farmsteads, has a history of light-touch settlement, albeit at times busier than today, but within this particular view it is forestry operations that dominate. If there is one place in the Lake District that demonstrates the turnaround in national forestry policy over recent decades, this must surely be it.

View from Castle How looking south

Your viewing position: Looking east across the river

Geology and hows: It is a curious fact that around Grey Friar, including its western slopes facing you, there is the densest concentration of hows in the Lake District. Geology is doubtless part of the reason; this is after all rugged volcanic country. The fellside opposite between Cockley Beck and Grey Friar is a recognised excursion for geologists to explore lower Borrowdale Volcanic Group rocks. Here the Geologists' Association field guide takes visitors to several hows when explaining the variety of geological features in the succession of andesitic and dacitic lavas interspersed with distinctive pyroclastic rocks in this area. However, the general craggy appearance of these fellsides is hardly unique within the volcanic areas of the District, so the exceptional number of features bestowed with the name "how" ought to have another explanation. But seeking such an explanation is complete guesswork. An historical Norse presence in the area can be assumed so one possibility is that in this remote location, unlike other areas, Norse-given names have survived better over the centuries. An alternative explanation might be that over the post-Norse centuries something in the community character or whims of local notables has sustained an exceptional enthusiasm for bestowing protruding landforms with names that reflect Norse heritage. A credible explanation is elusive.

Early history: Fieldwork, notably by a community archaeology project called Ring Cairns to Reservoirs, has revealed ring cairns, burial cairns and other features indicating a human presence in the Duddon valley from at least the Bronze Age. However, this far up the valley evidence of such early occupation has not been found, although it can probably be assumed that the area was visited in prehistoric times for hunting and summer grazing. During the Roman period a fort was built nearby in Eskdale (see the pages on Dod Knott) guarding the road from Ambleside to Ravenglass that came over the Wrynose Pass along Wrynose Bottom and then over the Hardknott Pass and down Eskdale. To your left the line of the modern road over the Hardknott Pass differs from that taken by the Romans. From Cockley Beck the Roman road ran to where Black Hall Farm is today and then wound its way up to the pass approximately along the line of today's public footpath. Somewhat surprisingly, there is no other evidence of Roman activity in the Duddon valley, which raises unanswered questions about how the garrison of such a remote fort was fed and supplied. After the Romans, the profusion of Old Norse place names indicates a Norse presence in the valley but tangible evidence of settlement, as so often in the District, is lacking. However, lower down the valley there is interesting evidence of longhouses of probable medieval age based on a typical Norse longhouse layout (see the pages on Castle How, Duddon).

Remains of Gaitscale farmstead

Medieval and post-medieval periods: The limited number of farmsteads suggests this upper valley was not a favoured location for manorial colonisation in the medieval period. The farmstead highest up the valley appears to have been at Gaitscale, to your left in Wrynose Bottom. It may have been based on a former Norse shieling established in the late 10th century and was occupied until the end of the 18th century. Here the remains of a farmstead and an enclosure of possible inbye land can be seen. Elsewhere there were farmsteads at Cockley Beck, Black Hall, Dale Head and Birks (left to right in your view). It is thought there were common fields on the valley floor between Cockley Beck and Dale Head, and the fells, which were part of the Forest of Millom, provided common grazing lands and rights of turbary (peat cutting), evidenced by Peat Hills marked on the Ordnance Survey map on the fellside opposite you. On and around Castle How there are a number of enclosure boundary walls, footings of buildings, clearance cairns and possible cultivated areas, of uncertain date in the medieval/post-medieval period.

In the 16th and 17th centuries, the inherent problems of the feudal system of land tenure (notably the constant subdivision through the generations) were addressed here as elsewhere in the District through the manor courts and petitions. Although some intaking would probably have already started in the late-medieval period, larger intake enclosures were built at this time, such as Cockley Beck Great Intake and a number of "closes" such as Dale Head Close, Old Close and Low Close opposite you. As part of this reorganisation, some farms were amalgamated or abandoned (such as at Gaitscale) and a number were rebuilt (such as Black Hall and Dale Head in the 17th century). Although there has been no 19th-century enclosure of the higher fell land, none of the unenclosed land in your view remains as common land today, with the exception of the land to your right on the south-west flanks of Grey Friar above Hinning House.

The only mining in this upper part of the valley was at two small copper mines at Cockley Beck. One, on the north side of Cockley Beck Gill, was operational around 1700 and the other on the south side is probably of 19th-century origin.

Hardknott Forest Park: Around you is a forest with a controversial history that is now in the process of transition. In 1934 the Forestry Commission acquired 7,000 acres (2,800 ha) of land in upper Eskdale and the upper Duddon valley for the creation of an intended "Hardknott Forest Park". When the Commission's plans became public in 1935 an enormous public outcry broke out with fears of a coniferous landscape like those that had been created in Ennerdale and at Whinlatter. A campaign was led by the Council for the Preservation of Rural England (CPRE), the newly created Friends of the Lake District (FLD) and a host of nationally prominent individuals. With different national and local perspectives, the amenity groups squabbled amongst themselves about priorities in their confrontation with the Forestry Commission but essentially their target was to minimise the area to be afforested in Eskdale and the Duddon valley and secure a state forestry exclusion zone for the central Lake District. In July 1936, after many difficult meetings between the parties, a written agreement was reached between CPRE (on behalf of the amenity groups) and the Forestry Commission on an exclusion zone covering 300 square miles (777 square km) of the central Lake District. This took in Blencathra in the north, Coniston Old Man in the south, High Street in the east and Pillar in the west. This extraordinary voluntary agreement between a major government agency and an amenity group is still in place today.

The boundary of the agreed forestry exclusion zone is a little complicated in the area around Castle How where you are, but essentially the north and west flanks of Grey Friar (acquired by the National Trust before 1936), the higher ground of Ulpha Fell and Hard Knott are within the exclusion zone, whilst Harter Fell, the west side of the Duddon valley up into Wrynose Bottom and the lower slopes of Mosedale are excluded. By the time of the 1936 agreement, the Forestry Commission had already started on its Hardknott Forest Park with planting on the land of the former Birks Farm on the flanks of Harter Fell. As the forest gradually extended during the late 1940s and 1950s, the FLD and the National Trust sought to persuade the Forestry Commission to give up its ownership of the lands of Black Hall Farm, which included much of Ulpha Fell, Mosedale and Hard Knott. Recognising that much of the land was actually unplantable because of poor soil conditions, the Forestry Commission agreed to sell not only Black Hall Farm but also Brotherilkeld Farm in Eskdale. In 1961, with the benefit of a bequest, the National Trust finally took over ownership of both farms. The Forestry Commission's full plan for the Hardknott Forest Park was abandoned; nevertheless, a substantial coniferous forest had been created, named Dunnerdale Forest on the Ordnance Survey map but generally referred to as "Hardknott Forest".

Restoring Hardknott Forest: Many centuries ago, all but the higher land you can see would have been forested; the Forestry Commission's aim now is to go as far as practicable in recreating this native forest. The forest originally planted in the 1940s and 1950s with spruce, larch and pine was felled and partially replanted in the 1990s but by the early years of this century the Forestry Commission and others had recognised an opportunity to take a new direction and create the largest semi-natural woodland in the Lake District, linking up with the long series of oak woodlands that run down the Duddon valley to Duddon Bridge (see pages on Castle How, Duddon). The aim now is to restore the complete forest into a mosaic of native habitats of oak and birch woodland, bogs, open woodland and rocky crags creating an ecologically and visually diverse environment. In the main, natural regeneration is relied on to create the new areas of native woodland with new planting only where local seed sources are not providing the desired new growth. Within the forest area there is no grazing of domestic livestock, but there is grazing by wild deer, which requires fencing off vulnerable areas. The Forestry Commission is assisted in this project by volunteer working parties and by university studies that are monitoring the woodland regeneration and the impact on ecosystem services, such as carbon storage, water quality, biodiversity and recreational use.

A volunteer removing a young non-native Sitka spruce

From where you are sitting on Castle How it is fairly obvious that you are looking at a forest in transition. Start by turning around and looking north-west, where you can see a corner of conifers that has been felled at the forest edge to break up a harsh straight-line boundary. Softening such plantation boundaries is a general objective across the forest, and where the natural seeding of new spruce outside their main blocks provides this benefit, they are allowed to remain; elsewhere young conifers are removed by the volunteer working parties. Then turning leftwards and looking down valley, the extensive new areas of regenerated woodland, largely birch and oak, can be seen surrounding older blocks of conifers. In the long term (post 2047), most of this view will comprise native deciduous woodland, interspersed with open areas. In the medium term, however, some of the stands of conifers, such as the conifer block in line with Harter Fell summit and the conifers beyond Pike How will remain until 2021–2026 to retain some structure in the appearance of the landscape. In future the newly regenerated areas will be managed through what is termed "continuous cover forestry", which involves the maintenance of the forest canopy and a presumption against clear felling. This is a form of forestry management that can be more expensive but brings a wider range of ecosystem services – that is, benefits for the environment and humans. Adopting this form of management is seen as key to meeting the objectives for the forest.

Dod Knott, Eskdale

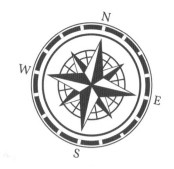

An excellent vantage point overlooking Eskdale near Brotherilkeld
with a stunning view of the valley from source to sea

Location:	NY209005; N54°23.604 W003°13.067
Height:	250 metres
Geology:	Borrowdale Volcanic Group
Ownership:	National Trust
Access:	Open access land; leave the bridleway at the gate at NY206002 and traverse the fellside over rough ground, closely following the fence that leads conveniently to Dod Knott
Website panorama:	From south-west through north to north-east

Dod Knott, right of centre, from across the valley

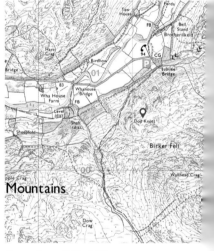

Location of Dod Knott

♀ The View

The valley scene below you lies on the threshold of the wildest part of the Lake District. This
is where becks cascading down from the dramatic, volcanic high fells combine to form a more
sedate river that winds through a sparsely settled, pink-hued granite landscape. Through this

valley the Romans built their road from Ravenglass to Ambleside, guarded by a fort alongside the steep climb over the Hardknott Pass. Later, the monks of Furness Abbey established the long tradition of Herdwick sheep farming in the valley and then, in the 1930s, these treeless tracts of upper Eskdale were purchased by the Forestry Commission prompting a pivotal clash with amenity groups and leading to a unique cash buy-out.

View from Dod Knott looking north

Your viewing position: Looking north-west across the valley.

Igneous landscape: This is a good spot to see the difference between the landscape of Eskdale granite, which lies across the river from where you are sitting and leftwards, and the landscape of volcanic rocks on your side of the valley and rightwards into upper Eskdale. Dod Knott itself is composed of dacite lava that was laid down in the early part of the volcanic period as a lava sheet lying almost horizontally, as can be seen fairly readily when viewed from the opposite side of the valley. These and other early Borrowdale Volcanic Group rocks extend leftwards to Birker Fell and on your right towards the high fells. The high fells themselves are formed of later complex pyroclastic and volcaniclastic rocks. Huge quantities of molten magma were intruded into these volcanic rocks at the end of the volcanic period 450 million years ago, cooling slowly to form the coarse-grained granites that create the distinctive Eskdale scenery. If you look across the river and leftwards you can see the typical bulbous granite outcrops on the valley sides and poorly drained low rolling moors above where the granite has responded to glaciation to produce numerous knolls and hollows. During glaciation ice moved south-west from the high ground of the Scafell area, with the ice from upper Eskdale and from Lingcove converging around Throstlegarth, and gouging out the deep trough to your right with its spectacular vertical buttress of Heron Crag and other valley-side rock features.

Prehistory and the Romans: Evidence of the presence of prehistoric peoples in the locality can be found on Eskdale Moor, the higher ground above the granite crags at a distance to your left. Here a number of stone circles and other features forming part of a ritual and funerary landscape, together with evidence of field systems and clearance cairns, suggest a thriving, most probably Bronze Age, farming community. During this period it was these slightly elevated locations rather than the heavily wooded wet valley bottoms that were most attractive for settlement. The area is littered with scheduled ancient monuments that are fascinating to explore.

Prehistoric stone circle on Eskdale Moor

Then more clearly within your view to your right is the spectacularly positioned Hardknott Roman Fort. It might not have appeared quite so spectacular when it was built because the evidence is that the area was wooded so, in all likelihood, the site would have required much tree felling ahead of construction. The fort was built in the early 2nd century AD during the reign of Emperor Hadrian with the primary purpose of protecting the road between the port of Ravenglass and the Roman fort at Ambleside. It was demilitarised around 140 AD when the Romans reoccupied southern Scotland and then re-garrisoned 30 years later before being finally abandoned in the early 3rd century. The fort is square with four gates and within the walls there are two adjacent granaries, a garrison headquarters and the commander's villa. Outside the fort there are the remains of a bathhouse and a fine example of a levelled parade ground some 200 metres away. Some of the buildings were subject to Victorian reconstruction with a slate layer marking the boundary between the original and the reconstruction. There is evidence that for a period the fort was garrisoned by the Fourth Cohort of Dalmations; one wonders what these soldiers from the sunny Adriatic made of this remote northern posting.

Eskdale farming landscape: In medieval times the pastoral farming regime along much of Eskdale differed from that at Brotherilkeld and the dalehead, which was part of the Furness Abbey estate. From where you are sitting you have a view of both landscapes. To your left along much the length of Eskdale peasants grew crops in the valley bottom and exercised their rights of common on the fells beyond the intake walls. These included rights for pasture, turbary (cutting peat for fuel) and estovers (cutting bracken for thatch and animal bedding). Over time there was some peasant enclosure and colonisation of waste (known as "assarting") that, although often technically illegal, provided the lord of the manor with useful additional income. Most of the farms in the valley today are on sites occupied since the 13th century. In the post-medieval period here, as elsewhere in the District, farming was reorganised largely on the initiative of the more entrepreneurial farmers, which led to farmsteads being amalgamated, rebuilt or abandoned. Today's farms are the survivors of this period with Wha House, Birdhow, Taw House and Brotherilkeld, all in front of you, rebuilt in stone in the 17th and 18th centuries. Although some new intake walls were built around the farmsteads, this part of Eskdale and surrounding fells were never subject to Parliamentary Enclosure.

Taw House Farm's particular claim to fame is as the welcome overnight refuge for the poet Samuel Taylor Coleridge after his nervy, but celebrated, descent of the challenging Broad Stand on Scafell in 1802. After breakfast, he was guided to what at the time was dubbed "Eskdale's most notable tourist attraction" – the "Four-foot Stone" at the foot of Buck Crag (High Scarth Crag), marked with human and animal footprints. Find it if you can.

Furness Abbey lands: In 1242 the existing dalehead farm of Brotherilkeld was granted to the Cistercian monks of Furness Abbey by the Lord of Millom (the Hudleston family). The Normans had introduced monasticism as an important part of the social fabric, with Furness Abbey, founded in 1127, one of a number of abbeys founded in Cumbria in the 12th century. Norman lords, particularly when land was not especially profitable, would choose to give land to monasteries as a worthy deed commensurate with their social status. Medieval documents describe the boundary of Furness Abbey's farm at Brotherilkeld and it is worth noting in view of what eventually happened to the land in the 20th century (described below):

from the Esk, down to your left, it ran up Spothow Gill to Harter Fell summit, then to Hard Knott, the head of Mosedale, Crinkle Crags, Bowfell, Esk Pike, Esk Hause and then down the Esk to complete the circuit below you. The Cistercians were good agriculturalists and improvers and by 1292 they had established a vaccary at Brotherilkeld. By 1536, at the dissolution of the monasteries, it had been developed into a large commercial Herdwick sheep farm, one of several sheep farms established by monastic orders that helped make the Lake District into an important centre for sheep farming and the wool industry. The Abbey's extensive grazing lands in upper Eskdale were adjoined on the west side of the Esk by the hunting grounds of the Forest of Copeland. The different uses of the land on either side of the Esk necessitated a formal agreement between the two landowners for the building of a dyke and wall that contained the sheep on the east side while still allowing wild deer and their fawns to leap over. Remnants of this dyke and wall can be seen on the Great Moss and are shown on the Ordnance Survey map.

Remains of the medieval boundary wall in upper Eskdale

Mining industry: Iron ore was mined and smelted in this part of Eskdale for hundreds of years. In the medieval period, iron ore was probably mined from surface veins and smelted in local bloomeries using locally produced charcoal. There are medieval bloomery remains just above Jubilee Bridge at the foot of the Hardknott Pass and a short distance up the Esk at Scale Bridge, and there are remains of numerous charcoal burning platforms immediately below Dod Knott on the far side of the bridleway and intake wall. In the latter part of the 19th century, the scale of the iron industry increased substantially, with a number of mines operating in the vicinity of Boot, and in 1875 a railway was opened between Dalegarth and the port at Ravenglass to transport ore out of the valley. The line was extended to Gill Force on the south side of the Esk in 1880. The railway had a very chequered commercial history into the 20th century carrying ore and granite, but is now transformed into a popular passenger-carrying tourist railway.

Hardknott Forest Park: In 1934, the Forestry Commission acquired 7,000 acres (2,800 ha) of land in upper Eskdale and the upper Duddon valley for the creation of the Hardknott Forest Park. In Eskdale the Commission had acquired Brotherilkeld Farm and all the land within the boundary described in the paragraph above. The planting that was proposed within this huge area, as part of the planned Forest Park, covered the lower north-western slopes of Harter Fell where you are sitting (from Spothow Gill round to Hardknott Gill) and the fellsides north-west of Hardknott Fort as far as Yew Crags. The remaining land in the Commission's ownership in upper Eskdale (the vast majority in fact) was deemed "unplantable", which naturally raised questions about why they had acquired it in the first place. Nevertheless, the land proposed for afforestation here in Eskdale, coupled with the planting proposed in the Duddon valley, provoked a very large campaign of objection. The campaign and the agreements eventually reached with the Forestry Commission in 1936 are summarised in the pages on Castle How, Upper Duddon.

One of the key agreements reached with the Forestry Commission was the creation of a state forestry exclusion zone covering the central Lake District. From your position on Dod Knott the section of the exclusion zone boundary lying within your view comes towards you from the direction of Wasdale, crosses the river Esk in the vicinity of the Scale Gill confluence and then goes up to Hard Knott. The area of upper Eskdale to the north of this line is excluded from future state afforestation, but the area of planned afforestation in 1936, described above, lay south of this line. As part of the agreements reached, the Forestry Commission had undertaken not to plant this latter area subject to the receipt of financial compensation at the rate of £2 per acre. This compensation had to be raised through a public appeal organised by the amenity groups. The required funds were raised and through an exchange of letters the Commission abandoned its Eskdale planting plans. This was meant to be formalised through restrictive covenants, but these were never drawn up, and then, sometime later in 1961, all the Forestry Commission land in upper Eskdale (that is Brotherilkeld Farm) was sold by the Government to the National Trust.

A footnote: You may be able to make out a new National Trust planting scheme of native tree species stretching up the valley from Brotherilkeld Farm to beyond Yew Crags. This scheme crosses the 1936 boundary but technically does not transgress the agreement because it is not a Forestry Commission scheme. Nevertheless, it is a sign of the new thinking about trees in the central Lake District.

Eller How, Winster Valley

A wonderful view over a mosaic of coastal pastures, woodland and low-lying mosses, overlooked by the dramatic limestone cliffs of Whitbarrow

Location:	SD414816; N54°13.576 W002°54.071
Height:	177 metres
Geology:	Windermere Supergroup
Ownership:	Private
Access:	Open access land
Website panorama:	From north through east to south

Eller How from the north

Location of Eller How

⚲ The View

Eller How, situated in the tract of Newton Fell known as Dixon Heights, provides a fine position to survey a landscape very different from typical Lake District scenery. It was into this low-lying, coastal landscape that the earliest groups of humans ventured and ultimately established a scattered, but continuous, presence to the present day. This is a landscape where human endeavour has wrestled land for productive agricultural use from its natural watery state, but is now the setting for skirmishes between different interests about how the water should be managed in the future.

View from Eller How looking north-east

Your viewing position: Looking north-east towards the cliffs of Whitbarrow Scar

Eller How folly: When you arrive on the top of Eller How you will see the ruins of a tower. It was probably built in 1845 as a folly by George Webster, a well-known Kendal architect who designed and lived in the elegant Regency villa, Eller How House, below.

Land shaping: Prominent in your view are the shapes of the upstanding north–south ridges of Whitbarrow and Yewbarrow. These are fault-bounded blocks of limestone which, together with other areas of limestone around Morecambe Bay, are part of the residual outer rim of the limestone originally laid down over the whole District. Most of the rest of the land within your view is underlain by older deep-sea mudstones and sandstones of the Silurian period, nearly all covered in deep glacial, fluvial and tidal-flat deposits. An exception is at the Holme, to your left, where the bedrock is exposed creating a rock barrier across the valley. This provided a dry route across the valley floor for travellers from earliest times and was the location of the proposed dam for an aborted Manchester Corporation reservoir scheme in 1964.

During the Ice Age there was a south-flowing glacier in the Winster valley and a larger one in the Cartmel valley to the west. During this time the Holme rock barrier was probably higher than it is today and had the effect of diverting some Winster ice south-west to High Newton where it joined the Cartmel glacier. Then during the immediate post-glacial period it is possible that the rock basin north of the Holme contained a temporary lake of which Helton Tarn is a remnant. As the ice melted at the end of the Ice Age, sea levels rose and flooded the Winster valley as far up as Coppy Beck, it is estimated, some six kilometres upstream from where you are. (This sea level rise did not reach the foot of the Whitbarrow cliffs, which mark a former coastline created during an earlier inter-glacial rise in sea level.) As the land mass, relieved of the weight of ice, gradually rose, the sea retreated from the Winster valley and a new landscape of open water, marsh, fen and raised peat bog evolved around the Kent estuary, with forest establishing itself on the slightly higher, drier ground away from the coast.

Early settlement: It was into this coastal landscape that some of the first groups of humans arrived around 10,000 years ago, attracted by the resources of fish, shellfish and wildfowl to supplement food gathering and hunting. Evidence of prehistoric presence comes from numerous finds of tools and arrowheads, the earliest being Mesolithic microliths found in nearby limestone caves. During the Iron Age there was a promontory hill fort at Castle Head, just south-east of Lindale, and various artefacts have been found there indicating Romano-British occupation. During the medieval and post-medieval periods there was some settlement in the Witherslack area, opposite you, with small farms growing oats and rearing sheep and cattle. The peat mosses were drained and cut for fuel and the valley's woodlands started to be exploited for charcoal production and iron smelting.

Enclosure: In the late 18th and early 19th centuries the Enclosure Acts did much to change the farming landscape of the Winster valley. Newton Fell and the land west of the Winster were in the parish of Cartmel (at that time part of Lancashire) and were subject to the provisions of an Act of 1796 to enclose the Cartmel commons, waste grounds and mosses. Land on the east side of the Winster was in Westmorland and subject to the Witherslack Enclosure Act of 1815. The Annals of Cartmel, published in 1872 by James Stockdale, a local landowner, give a good insight into the work of Enclosure Commissioners and the impact of enclosures in the area. The Annals report that the Cartmel parish enclosures were "a great and memorable event, causing as it certainly did the conversion of an extensive tract... of bare, because always overstocked, pasture land, used in common, into numerous well-fenced enclosures, more or less capable of cultivation, set out and allotted by the Commissioners". The allotments were made by the Commissioners to existing claiming occupiers "in severalty" (giving sole ownership) subject to certain qualifying criteria. In addition, the enclosure procedure commissioned the creation of roads with surfaces "covered in small broken stones... in place of ruinous old clog-wheel-cart and pack horse tracks", the building of wider bridges and most significantly the digging of "several deep main drains cut through swamps, mosses and low grounds". The idea of the enclosures and associated new infrastructure was primarily to provide an incentive for agricultural improvement, but the Annals also report a much-appreciated decrease in the "ague" amongst the population, arising from the drainage of stagnant sheets of water.

Tracks and roads: Travelling through the low-lying coastal landscape in front of you was always a serious challenge. From earliest times tracks would have existed, some constructed as "corduroy" roads over the peat with logs laid perpendicular to the direction of the road, supported and staked by other logs. However, by medieval times these local ways were not a practicable means of communication as trading and administrative links built up between Kendal, Lancaster, Cartmel and Furness. Transport of goods was commonly by sea via the shallow waters of the Kent estuary and travellers would take a coach over the Morecambe Bay sands. Land routes between Kendal and Furness took routes some distance inland via Bowland Bridge and over the Holme. Half-decent road communication to Furness was finally established in 1818 when a new turnpike road was opened roughly along the alignment of the present A590. It proved a challenge to build, with sections over the mosses reportedly floated on bundles of juniper cut from the slopes of Whitbarrow. Even today dips appear periodically in the modern dual carriageway, demonstrating its less-than-solid foundations.

Whitbarrow SSSI: In front of you across the Winster, the two prominent limestone ridges of Whitbarrow and Yewbarrow and the woodland immediately below them are of outstanding nature conservation interest. The area is designated a Site of Special Scientific Interest and is part of the Morecambe Bay Pavements Special Area of Conservation. The site is a complex mix of limestone and acidic grasslands, heath, scree, cliff, limestone pavement and woodland providing an extensive and varied series of semi-natural habitats. The area supports significant populations of nationally rare plants and notable invertebrate fauna. Historically, the area provided valuable grazing for sheep and cattle before the draining of the low-lying surrounding lands for pasture in the 18th and 19th centuries. The woodlands were coppiced and provided a range of wood products, and limestone was quarried for building stone and burnt in nearby limekilns to make agricultural lime.

Flooded Winster valley and Witherslack Mosses in December 2015

Today the area is owned and managed by a number of private, public and charitable organisations with a mix of both commercial and nature-conservation objectives. Grassland areas provide grazing for cattle and sheep, which prevents the expansion of bracken and woodland and increases the number of flower species present. Some of the woodland is managed as coppice, which encourages violets and primroses that in turn attract butterflies, and in recent years areas of planted pines have been removed to allow native woodland and open grassland to regenerate.

Ancient woodland: Nearly all the woodland you can see looking towards Whitbarrow is categorised as ancient semi-natural woodland with some large planted ancient woodland sites; these are woods of tremendous historical and ecological interest. In the medieval period deer, birds and other animals were hunted (there is documentary reference in 1341 to a deer park in the woodland below the left-hand skyline of Whitbarrow) and the lord's tenants would have had rights to graze animals and take specified categories of wood for fuel, building and other purposes. By the 18th century the lord's tenants, as well as using it for their own purposes, were trading their wood resources to make charcoal for iron manufacture and with wood craftsmen, both local and from outside the area. Records from 1786 reveal that in the parish of Witherslack there were a range of woodland-dependant occupations – coopers, turners, carpenters, box makers, basket makers and charcoal burners. Today most of the woodland you can see is part of the Witherslack Estate, land owned by the Stanley family (the Earls of Derby) since Tudor times. A couple of decades ago the commercial prospects of privately owned broadleaved woodlands would not have been rated very highly, but today the value of these woodland habitats is better appreciated and a combination of new markets and grants has turned the situation around. The growth in demand for wood fuel has been substantial in recent years and this has underpinned a programme of woodland regeneration here at Witherslack, including bringing traditionally coppiced woodlands back into rotation.

Witherslack Mosses: In front of you the Foulshaw, Meathop and Nichols mosses are the sole survivors of a large wetland that began to form around 6,000 years ago after the sea had retreated following the sea level rise at the end of the Ice Age. Within the wetland, extensive domes of peat rising above the surrounding land were formed from waterlogged sphagnum moss building up layer by layer over thousands of years. However, much of the mosses have been lost to peat cutting and land drainage for agricultural purposes. One of the contributors to this was John Wilkinson, a famous and very enterprising ironmaster who lived at Lindale for a number of years at the end of the 18th century and is buried there. He had the idea of using peat rather than charcoal to smelt haematite ore. It didn't work out but it did lead to Wilkinson becoming a pioneer in techniques to bring mosses into agricultural use through land drainage and other improvement techniques. The raised peat bogs that remain today, collectively called the Witherslack Mosses, are now examples of one of England's most scarce habitats. This habitat with its colourful patchwork of sphagnum is important for many species of rare insects, flowering plants and birds. In the past, forestry, drainage and scrub invasion on these sites created a harmful dry and shaded environment but now a long-term programme of restoration of the mosses is in progress. Trees have been removed and drainage channels have been blocked to raise water levels so that the process of recreating the natural sponge characteristics of a peat bog can get underway. These three raised peat bogs are Cumbria Wildlife Trust reserves and are now managed solely for nature conservation and educational purposes.

Peat cutting at Foulshaw Moss; early / mid 20ᵗʰ century

Water management: Farming and living in this landscape have always required wrestling with water levels. With climate change predicted to raise sea levels and bring more river flood events, this coastal area will be fending off water from both seaward and landward. What to do, or not to do, with respect to threats from the sea is currently being looked at as part of the preparation of an updated Cumbria Coastal Strategy being coordinated by Cumbria County Council. On the landward side, much of the agricultural land you can see around Witherslack and further right beyond the A590 to the estuary is kept drained through maintained drainage channels and pumping stations. In 2011 the Environment Agency notified land occupiers that it would cease to operate pumping stations in the neighbouring Lyth valley and at Ulpha (to your right, adjacent to Crag Wood by the estuary). This decision was in line with the Agency's policy to focus its resources on alleviating flood risk to property and people, but not to agricultural land. This notice of intent caused consternation amongst the farming community, but in the minds of conservation bodies opened up opportunities for re-establishing natural processes and enhancing biodiversity – a stark clash of visions about the future of this landscape. At the time of writing, implementation of the Environment Agency decision has been delayed and the possibility of setting up a Lyth and Witherslack Water Level Management Board is being explored.

Gummer's How, Windermere

Sensational panoramic vistas of the southern Lake District combined with closer views of Windermere and its shores

Location:	SD391885; N54°17.303 W002°56.341
Height:	321 metres
Geology:	Windermere Supergroup
Ownership:	Private
Access:	Open access land; a footpath leads easily to the summit from a small Forestry Commission car park on Fell Foot Brow
Website panorama:	From south through west to north

Gummer's How from Lakeside

Location of Gummer's How

The View

After taking in the wonder of the 360° panorama, what holds the eye is the long ribbon lake of Windermere stretching north as far as you can see; a former highway for the transport of goods, the lake is now plied by tourist boats and enjoyed by recreational sailors. The land rising gently from both lake shores has a parkland character with extensive deciduous woodland interspersed with open areas and occasional large houses. This is a view shaped by medieval monasteries, woodland industries, a railway company, wealthy landowners and tourism entrepreneurs to create a mosaic of interest within a restful lakeside setting.

View from Gummer's How looking south-west

Your viewing position: Looking west across the lake

Windermere and its outflow: This low-lying, undulating country, based on the rocks of the Windermere Supergroup, contrasts sharply with the higher and older volcanic rocks of the central Lake District that can be seen in the far distance to your right. The shores of Windermere cut across the Silurian slates, mudstones, sandstones and siltstones, and you can see where harder beds stand out as promontories and weaker ones as bays. Windermere, in a trough carved out by a large glacier, is a classic ribbon lake with a maximum depth in this south basin of 40 metres, just off Grubbins Point to your right. The depth of glacial and alluvial deposits on the lake floor hides the true depth of the excavated rock basin, which by some recent estimates could be around 120 metres below sea level.

If you look at the outflow from Windermere (the River Leven) to your left, you will see that it turns sharply south-west rather than continuing directly south down the Cartmel valley, the main line taken by the Windermere glacier. In the past it was thought that this redirection occurred because the post-glacial lake had been dammed by moraines that diverted water over a low col south-westwards. This theory did, however, raise the question of how the deep and narrow Leven valley could have been eroded in the short post-glacial period. So, in 1986, seismic surveys were conducted and revealed that three kilometres due south of the end of the lake, in the vicinity of Fiddler Hall Wood, there is a low "saddle" of bedrock beneath the glacial drift at a height of 60 metres above sea level, some 21 metres above the present surface level of Windermere. It is the rock barrier at this height and position, not glacial drift, that diverted the outflow down the Leven valley, now thought to have been partially formed prior to the Ice Age. The rock barrier near Fiddler Hall Wood is in reality the south lip of the Windermere south basin.

Medieval land management: Information about early settlement on the east and west sides of this part of Windermere is very sketchy. It is only with the establishment of the Norman monastic houses that we gain some idea of how the landscape was used and evolved. The land opposite you, between Coniston Water and Windermere, known as High Furness, was granted to the Cistercian monks of Furness Abbey in 1127 and the lands on the east side of the lake (as far as Ghyll Head, four kilometres north) were an extension of the Cartmel valley lands granted to the Augustinian canons of Cartmel Priory shortly after 1189. The two religious houses, based on their respective religious traditions, managed their lands in different ways. The Cistercians believed in self-sufficiency and, with the support of lay brothers, did much of the physical farming work in producing food crops and rearing cattle and sheep. An essentially pastoral landscape was created in High Furness, with unenclosed rough grazing lands and woodland interspersed with farmsteads, often referred to as "granges", within their cleared enclosed lands. Across the lake it is thought there were granges at Stott Park and Finsthwaite, slightly to your left, and at Graythwaite, further away to your right. By 1292 the Abbey's wider interest in the iron industry becomes apparent with iron-making bloomeries operating across High Furness, fuelled by charcoal made from the local wood supplies. On the east side of the lake, the primary vocation of the canons of Cartmel Priory was pastoral care and providing spiritual services to the local people who were their customary tenants, farming what was mostly common land. The Augustinians' income from their spiritual services and their tenants meant they were not as involved as the Cistercians in the direct management of agricultural land. As a consequence, few granges were established on Cartmel lands and none at all on the eastern side of the lake (the closest being to the south where Grange-over-Sands now lies).

Past woodland management: By the time of the dissolution of the monasteries in 1536, the woodlands that are so important in the scene you are looking at today were in a state of serious decline. They provided wood for many wood-based industries, but in particular they were being decimated to make charcoal to fuel the numerous iron bloomeries located in High Furness and on the east side of the lake. The evidence of bloomeries and charcoal pitsteads can be found in many of the woods across the lake and around you. Iron ore, easier to transport than charcoal, was brought in by pack horse and boat from the rich haematite pits in the limestone areas near Dalton in Low Furness.

A wood stack being covered with turf and grass in preparation for firing to make charcoal; location in south Cumbria in 1900

By the latter part of the 18th century the increasing demand for wood was prompting a reversal of woodland loss with the planting of new timber and coppice plantations. The coppice plantations were a particular feature of High Furness providing, as well as charcoal, wood products for a host of other industries such as tanning, mining, shipping and textiles as well as for building construction and furniture making. Some thinking about sustainable woodland management had arrived. This was just as well because by the latter part of the 18th century, blast furnace technology had raised the demand for charcoal even further. Two such blast furnaces at the time were at Cunsey Beck, to your right across the lake just beyond the prominent Rawlinson Nab, and to your left down at Backbarrow, beyond Newby Bridge. While the demand for charcoal was an important incentive for investment in new coppice plantations, there were also rising demands for other wood products that encouraged investment in timber plantations. One investor was Richard Watson, the Bishop of Llandaff, Cardiff, and a native of Westmorland. He acquired 784 acres (317 ha) of common land at Gummer's How as part of the 1796 enclosures of the Cartmel Commons and planted a substantial larch plantation. The early tourists in the District commonly made reference, not always favourable, to the impact these new plantations had on the landscape, but they did nevertheless redress the catastrophic earlier woodland losses, and did much to create today's attractive wooded scenes. (Note the common naming of woodland as "plantation" on the Ordnance Survey map in this area.)

Woodland management today: Nearly all the woodlands on the slopes opposite (down to the lakeshore) and below you (between the A592 and the lake) are categorised as either ancient semi-natural woodland or planted ancient woodland sites. Most of these woods are privately owned and in recent years there has been a resurgence of interest in their management due to more profitable markets for wood products and publicly funded financial incentives and advisory services. These woodland owners are attracted by a combination of the commercial markets for various timber products and, more recently, for wood as a green biofuel, as well as the opportunities for improving amenity and nature conservation. Nearly all the private woodlands you can see have benefitted in recent years from publicly funded grants that aim to assist land managers in protecting, improving and expanding their woodlands, enhancing biodiversity and combating tree diseases. While you are on Gummer's How, you may see the Luing cattle that have been brought on to the land through a grant scheme to enhance biodiversity.

Newby Bridge and Lakeside: The developed waterside areas below and above Newby Bridge, down to your left, were historically a busy commercial transport hub where the movement of people and goods north–south along Windermere and the Leven valley intersected with the east–west land routes between Kendal and Furness. Originally the medieval route from Kendal ran westwards to Bowland Bridge, down Fell Foot Brow and down to the lakeside where there were fords at two stony shallows, one at Fell Foot and another a little further downstream at the Landing. Newby Bridge was the point where goods (such as slate, iron ore, coal, charcoal, stone) transported by sailing barges up and down Windermere were loaded and off-loaded. The fast waters of the River Leven downstream of Newby Bridge powered numerous industries over the centuries: bobbin, woollen, cotton and paper mills, the large Backbarrow charcoal-using iron works and the "Dolly Blue" ultramarine pigment factory.

During the early 19th century, a scheduled sailing-boat service for passengers and general goods operated between Ambleside and Newby Bridge. Then in 1845 the first steam-driven boat, *Lady of the Lake*, was launched at Newby Bridge with much ceremony. Built at Greenodd, the boat had to be designed as a paddle steamer because of the very shallow waters at Newby Bridge. Even so, a channel 1.5 metres deep was dredged northwards to ensure safe passage and this finally brought to an end the two ancient fording points. (The water level of Windermere did not rise to a height close to the present level until the weir just downstream of Newby Bridge was built in the 1930s.) The growing number of visitors seeking steamer trips on Windermere soon attracted the attention of the Furness Railway Company, which decided to build a railway spur up the Leven valley to the lake; but rather than terminating at Newby Bridge, they chose to build a terminus and pier by the deeper waters at a location they named Lakeside. The line opened in 1869 and from this time on Lakeside took over from Newby Bridge as the southern terminus for boats on Windermere. Although freight (such as coal for the steamers and iron ore for the Backbarrow ironworks) was initially the main income stream for the railway, by the end of the century the passenger business was thriving. However, the increasingly widespread use of the motor car during the 20th century brought a drop in passenger numbers and the line was closed in 1967. In 1973, part of the line reopened as a steam-hauled heritage railway.

Grand houses and hotels: From the late 18th century and during the 19th, the scenic qualities of these southern reaches of Windermere and the terminus for the steamers led to modest hostelries becoming luxurious hotels and farmhouses becoming grand residences. From the late 17th century, there was a coaching inn at Lakeside that was much used by travellers to the local woodland industries; but after the building of the railway terminus and pier, the Lakeside New Hotel was built, around 1875.

Lithograph of the former Fell Foot villa

The Swan Hotel at Newby Bridge, built in the late 18th century, was originally a posting house on the Kendal to Ulverston turnpike road. In 1784, a yeoman farmer's land and farmhouse at Fell Foot (down to your left) was sold to Jeremiah Dixon, a Leeds merchant, who then built one of the first Lake District villas, together with designed gardens. A much later owner demolished the main house in 1907 with a view to rebuilding, but this project had to be abandoned for family reasons. The land was donated to the National Trust in 1948 and is now a popular recreational and boating centre. Across the lake to your right is Graythwaite Hall, a 17th-century house surrounded by parkland on one of the largest estates in the District. For 500 years it has been run by the Sandys family, who in 1886 called on the up-and-coming landscape designer Thomas Mawson to redesign the gardens in the Arts and Crafts style. This set him up for commissions elsewhere on the shores of Windermere (see pages on Allen Knott).

High Buck How, Borrowdale

A very fine viewpoint overlooking an enclosed valley landscape from which escape is only possible over high passes or through the "Jaws of Borrowdale"

Location:	NY258131; N54°30.483 W003°08.860
Height:	450 metres
Geology:	Borrowdale Volcanic Group
Ownership:	Private
Access:	Open access land; either ascend from the path south-west of the campsite or descend north from Bessyboot
Website panorama:	From south-west through west to north

High Buck How, left skyline high point, from the roadside

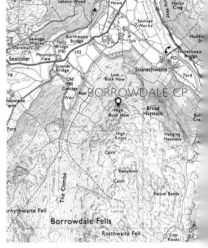

Location of High Buck How

📍 The View

Below is a view of green pastures, craggy fellside, rivers, tumbling waterfalls, oak woodlands, small hamlets and farms: a Lake District upland pastoral scene par excellence. This is a landscape shaped by the abbeys of Furness and Fountains and then taken in hand by the yeoman farmers after the medieval period. Farming has carried on stoically despite regular deluges of floodwater sent down from the wettest spot in England. Quarrying slate and mining graphite supported local incomes over several centuries, while woodland industries decimated the local woodlands until they were eventually replanted to give us today some of Lakeland's finest oakwoods.

View from High Buck How looking north

Your viewing position: Looking north-west over High Doat towards High Spy

Jaws of Borrowdale: Look to your right and you can see that the valley downstream is almost completely closed off at a point marked by the prominent rocky summit of Castle Crag (the site of an Iron Age fort and one of Thomas West's viewing stations). This constriction gained the name "the Jaws of Borrowdale" from early visitors to the valley. It marks the point at which, during the Ice Age, the erosive power of the valley's north-flowing glacier was seriously blunted by exceptionally resistant volcanic rock and eroded no more than the narrow gorge through which the River Derwent flows. At times, the glacier largely filled the valley; opposite you at the back of High Doat and at the back of Castle Crag there

Ice-marginal channel at High Doat (foreground) and at Castle Crag

are shallow valleys (followed by the present-day bridleway) that are thought to be the line of an ice-marginal channel where a stream ran along the upper edge of the glacier. For a period at the end of the Ice Age, the constriction at Castle Crag impounded water in a shallow temporary lake that may have extended some distance up towards Seathwaite until the present gorge was fully formed.

Glaciers' advances and retreats: Below you on the valley floor are a number of moraines that have for years been discussed by geomorphologists trying to interpret what they tell us about the comings and goings of the valley's glaciers. Moraines are ridges or mounds of glacial debris deposited in different settings, including at a glacier snout. Moraines at this location reflect the shape of the snout and are broadly arcurate in form with the concave side facing up-ice. If you look down in front of you, immediately left of the camp site, you should be able to make out the line of a moraine curving in the direction of Rosthwaite, with another one a little further left (both visible on the photograph above and on the website panorama). The concave side of the moraines reflects the shape of the glacier snout and this tells us they were deposited by a glacier in the Stonethwaite valley, not by a glacier in the Seathwaite valley. There has been debate amongst geomorphologists about when these moraines were deposited and current thinking is that they may be the product of a more recent glacier that extended from a plateau icefield that became established on High Raise and Thunacar Knott during the short Younger Dryas cold period.

Settlement history: Until recently there was no physical evidence of settlement in this part of Borrowdale before the medieval period. However, recent research commissioned by the National Trust has found field evidence of early settlement to the south-east of Thorneythwaite Farm, down to your left where four field walls intersect. A well-defined enclosure bank has been identified, some subdivisional walling and possible structural remnants. The site has not yet been excavated, but is thought to be of Iron Age/Romano-

British date suggesting that this might be the location of the community that used the fort on Castle Crag as a place of refuge.

The Manor of Borrowdale was established following the Norman arrival in Cumbria in 1092. A hundred years later, Alice de Rumeli, the grand-daughter of the manor's first Norman lord, granted her lands in Borrowdale, including most of the land within your view, to the Cistercian monks of Fountains and Furness Abbeys. The monks cleared and drained land, brought it into cultivation, coppiced the woodland and turned huge tracts of the fells into sheep pastures grazed by large flocks of sheep. The production of wool was the main business together with cattle rearing and the cultivation of rye, oats and barley. At Stonethwaite, round to your right, Fountains Abbey established a vaccary and the monks of Furness Abbey, which held the lands in the main Borrowdale valley south of Derwent Water, established a grange for the storage of farm produce; this is now the site of the present-day hamlet Grange. Sheep farming was developed on both monastic estates and by the 15th century there were dozens of farmsteads in this part of the valley indicating a healthy economy and substantial population. Following the dissolution of the monasteries, the Furness Abbey lands were retained under the control of the Crown until 1613 when James I sold them to two London entrepreneurs. They in turn sold them (except the valuable graphite mines at Seathwaite, discussed below) in a series of transactions, including through an agreement known as the 1615 Great Deed of Borrowdale, to a group of 38 farmers and gentlemen, headed by Sir Wilfred Lawson who became the dominant landowner. This had the effect of establishing a band of yeoman farmers in this part of the valley with the confidence to rebuild and invest in their buildings and land.

Land enclosures: By 1700, the valley floor below you had an established enclosed field system and there was some intaking up the valley sides, but tracking the earlier history of field enclosures as a way of understanding the land-use history has proved elusive. There is no documentary evidence of a medieval valley ring garth in upper Borrowdale and the evidence on the ground is very sketchy indeed, mainly because regular valley flooding and subsequent wall rebuilding have masked whatever evidence there might have been. Generally, within this valley head the ancient walls have been lost. Documentary evidence suggests that, by the time of the dissolution, parcels of land associated with individual farms were enclosed with small common fields at the hamlets of Seathwaite, Thorneythwaite, Seatoller and Rosthwaite. Documents of 1678 refer to "closes" at Rosthwaite, which may refer to enclosure of the strips in the common field. If you look to your right, the pattern of the walls to the right of the hamlet appears to reflect the pattern of old strip fields. The intake walls in this part of the valley appear to have been in place by the 1750s and by 1850 changes and additions to the pattern of walls within the valley had come to an end. The pattern of walls you look over today has not changed since then.

Wad and pencils: To your left, on the fellside to the right of the hanging valley of Sourmilk Gill above Seathwaite, is the birthplace of the world-famous Cumberland pencil industry. Here an exceptionally rare resource of pure graphite or "wad" was mined for 500 years. Initially the wad was used for medicinal purposes, marking sheep and dyeing cloth, and later for casting cannon balls and other smelting processes due to its ability to tolerate high temperatures, and then in the manufacture of pencils. Its versatility made it such an enormously valuable resource that it was a prime target for thieves. An Act of Parliament in 1752 tried to give protection to the mines, including through the right to mount armed guards, but inevitably smuggling activities were difficult to control in such a remote location. The first pencil factory opened in Keswick in 1832 and developed into the Cumberland pencil industry, which exported across the world. The mines, worked spasmodically throughout their history, were finally abandoned in 1891. The site is now a scheduled ancient monument.

Inside the Seathwaite graphite mine

Honister slate mines: Looking slightly to your left, you can see the valley leading up to the Honister Pass and the Honister slate mines. The working of the attractive Westmorland green slate got underway in the 17th century and, as the demand for roofing slates grew, reached peak production in the latter part of the 19th century. The slate was taken by horse and cart to Keswick for export out of the District. This was a cumbersome mode of transport that led to a proposal in 1882 to construct a railway from the mines to the railway at Keswick. The proposed route for the railway would have come down from Honister Pass towards you and then followed the line of the present-day bridleway behind High Doat, then down Borrowdale and along the west shore of Derwent Water. The Parliamentary Bill seeking approval of the scheme was withdrawn following a campaign of opposition led by Canon Rawnsley, the vicar of Crosthwaite in Keswick, later one of the founding members of the National Trust.

Borrowdale's woodland: From where you are sitting you have a view of some very fine deciduous woodlands, nearly all owned by the National Trust. Much of it is part of the Borrowdale Woodland Complex, an internationally designated Special Area of Conservation. This woodland is valued as part of one of the largest upland oak woodlands in England, especially rich in lichens, mosses and other rare plants. This ecological richness and apparent maturity belies the fact that most of these woods were planted only a century or so ago following their depletion during earlier centuries to make, in particular, vast quantities of charcoal. The remains of large numbers of charcoal pitsteads have been found in Johnny Wood opposite you, in the woods at the foot of Thornythwaite Fell down to your left and in Frith Wood just beyond Rosthwaite. Four medieval charcoal-using bloomery sites for the smelting of iron ore have been identified in Langstrath. These were operated by Fountains Abbey, sourcing the iron ore from locations such as Ore Gap between Esk Pike and Bowfell. Today all these woodlands are under active management by the National Trust with the objectives of increasing the extent and improving the structure of the woodlands, controlling grazing, creating a more diverse ground flora and associated wildlife, improving the visitor experience and generating income from timber and wood fuel.

Thorneythwaite Farm: In 2016 a major controversy arose over the sale of Thorneythwaite Farm on the valley floor a little to your left. The farm's land was the inbye land north and south of the farm buildings and much of Thorneythwaite Fell to your left. It had been identifiable as an independent fell farm for 1,000 years, but in 2016 it was without a tenant and the owner decided to sell. It was auctioned in three lots in this order: (1) the farm buildings and curtilage, (2) the rest of the farm's land and flock of 470 hefted Herdwicks and (3) the whole farm with a starting price of bids 1 and 2 combined. At the auction, the farm buildings and curtilage were sold to a private interest. Then the farm's land and sheep were sold to the National Trust (the neighbouring landowner) with a bid so extraordinarily high that it took everyone's breath away. Unsurprisingly, there was then no interest in the combined lot for the whole farm. As a consequence, the farm was broken up and the National Trust was widely castigated for betraying the trust placed in it by Beatrix Potter and others to work to sustain the integrity and viability of the District's Herdwick sheep farms. The Trust defended itself by saying it saw great benefit in the farm's land coming under its stewardship as part of its Borrowdale estate but did not have sufficient resources in place to make a bid for the whole farm that it could be confident would be successful. Wrapped up in this controversy lies one of the central dilemmas in managing the Lake District landscape today: how to manage a landscape holistically in a way that meets both the present environmental challenges and secures the viability of upland sheep farming.

Hutching's How, Wasdale Head

A quirky but well-placed viewpoint in dramatic surroundings
overlooking Wasdale Head's famous walled fields

Location:	NY192084; N54°27.849 W003°14.857
Height:	130 metres
Geology:	Borrowdale Volcanic Group
Ownership:	National Trust
Access:	Open access land; from the footbridge and public footpath over Lingmell Beck, follow a good sheep track northwards on the upslope side of the wall until below Hutching's How; climb up directly
Website panorama:	From south through west to north

Hutching's How, centre, from the south-west

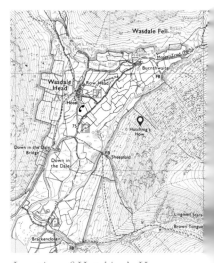

Location of Hutching's How

The View

This dalehead is surrounded by some of the finest wild scenery in England. The view from
Hutching's How overlooks the confluence of two dramatic glaciated valleys that lead down
to Wast Water and its spectacular screes. During the medieval period and later centuries

the most distinctive farmed landscape in the central Lake District evolved here, providing one of the District's most eye-catching scenes. In the late 19th century, visitors came – not searchers for the Picturesque, but educated, well-to-do men seeking the challenges of wild places. Through their endeavours and adventurous spirit, Wasdale Head became a famous pioneering centre for the new sport of rock climbing. Today it is the taking off point for thousands of walkers and runners seeking the summit of Scafell Pike, England's highest mountain.

View from Hutching's How looking north-west

Your viewing position: Looking north-west into Mosedale

A glacial confluence: The impressive scene you are looking over was fashioned out of volcanic rocks at the confluence of two valley glaciers during the Ice Age. Directly in front of you, looking up Mosedale, you can see the classic cross-section of a glacial trough; this was largely, but not entirely, shaped by a glacier that joined a larger glacier flowing right to left in the main valley now containing Lingmell Beck. The floor of Mosedale is at a slightly higher level than the main valley as a consequence of the difference in erosive power between the two glaciers, leaving a "hanging valley" down which Mosedale Beck tumbles in a waterfall named "Ritson's Force" (more about Mr Ritson later). To your left, tributary ice flows from the valleys containing Lingmell Gill, Over Beck and Nether Beck joined the main valley glacier to gouge out the deep basin that now contains Wast Water, a typical glacial ribbon lake. This is England's deepest lake with a maximum depth of 76 metres, some 15 metres below sea level (the depth of the underlying rock basin is unknown).

Glacial aftermath: After the glaciers retreated, vast quantities of glacial till were left behind covering the valley floor and part way up the valley flanks, softening any resemblance to the textbook 'U'-shaped glaciated valley. During the tundra-like post-glacial period, scree eroded from high crags covered the higher areas of till and on the valley floor becks have reworked the till, been realigned and deposited layers of alluvium that became the foundation for soils. Away to your left, the spectacular Wast Water screes are also the product of post-glacial natural processes. When the glacier melted, it left behind a rock landscape that was unstable through lack of ice support. Whole crags collapsed and rock-slopes shifted in several places along the Illgill Head ridge leading to huge rockfalls that accumulated on the slopes below. These dramatic processes are thought to be largely responsible for the creation of the screes; more commonplace talus eroded from cliffs since the Ice Age has simply added to the pile.

Settlement decline: Today the scatter of buildings across the valley floor in front of you belies a more substantial agricultural community that existed in the late-medieval period. The earliest documentary evidence of permanent settlement is from 1322 when there were three vaccaries established by the lord of the manor and run by tenants-at-will (tenants without security of tenure). The vaccaries are thought to have been at Burnthwaite, across the valley floor to your right, The Row, opposite you around where the Wasdale Head Inn is today, and at Down in the Dale further to the left. These vaccaries evolved and were subdivided over the 14th and 15th centuries into tenanted farms on which sheep were also raised. Records from the 16th century, when the wool trade was going strong, indicate that there were eighteen farms – four at Burnthwaite, eight at The Row and six at Down in the Dale – all grouped on the edge of the common open field on the valley floor with the tiny church, there in 1552 if not before, central to the scene. Adding to the busy scene would have been the packhorse trade coming and going over Styhead and Black Sail passes linking the central Lakeland valleys to the coast.

By the end of the 16th century, population growth in the Lake District had slowed and the farming practices of the yeomen farmers were holding sway. At Wasdale Head the common field was beginning to be reorganised and enclosed largely on the farmers' initiative, and the manor courts were directing the tenants' use of the huge areas of surrounding manorial waste by reorganising the management of the fellsides, moors and felltops. The most valuable lower fellsides (referred to as "banks") were reserved as good-quality cow pastures and the moors, like Burnmoor just out of sight to your left, were reserved as pasture for horses, heifers, bullocks and dry cows. The high fells encircling the main and tributary valley heads were reserved for sheep grazing in summer and divided into seven heafs, each assigned to particular farms.

By the latter part of the 18th century, economic and land ownership changes had led to the amalgamation of farms and abandonment of many cottages. The 18 farms of the 16th century had been reduced to eight by the end of the 18th century and a visitor in 1795 noted "vestiges of many ruined cottages show that this village was once more considerable". Today there are only three farms remaining, all owned by the National Trust; a solitary field barn and the footings of abandoned buildings are the only reminders of the settlement of Down in the Dale.

Wasdale Head school in 1890

The Wasdale Head walls: The National Trust owns most of the land you can see and has undertaken surveys that give a good picture of landscape history, including the building of the valley floor and other walls. The first stage of wall building on the valley floor was a ring garth enclosing the common field, which stretched from the head of Wast Water to a little above Burnthwaite farm. At first a small area of the slightly higher land at Burnthwaite was enclosed, possibly in the 11th or 12th century. The ring garth had been completed by the 13th century, possibly prompted by the farming necessities of the three vaccaries. The subdivision of the common field got underway in the early 17th century with the area in front and to your right enclosed by walls and the less-well-drained area to your left by "kesks" (hedgebanks). The stone walls display a fascinating, complex pattern much photographed from afar. In detail, the walls display surprising variations in thickness –

some have clearance cairns abutting them, some have been thickened along whole lengths ("consumption walls") and others infilled at wall corners. The complex layout of the walls is thought to have been determined by a combination of the boundary of the first enclosure at Burnthwaite, the lines of the original field strips (sometimes subdivided) and the courses followed by now relict streams across the valley floor. The stone for building the walls was clearly a case of resource availability exceeding requirements and originated from the twin sources of ploughing and deposition by Lingmell Beck and fellside becks when in spate. Just outside the northern and western boundaries of the open field, you can see the piecemeal enclosure of parcels of low fell land as intakes that took place over an uncertain period, but were largely complete by the end of the 18th century.

Other enclosures: Other walls to the left of where you are sitting have origins in two deer parks. There are historical references from the 16th century to "fences" enclosing land stretching up onto Illgill Head (above the Wast Water screes) that are thought to refer to two contiguous deer parks with origins at different dates during the late-medieval period. They were used by the lord of the manor before becoming tenanted land at a later date. Present-day walls follow much of the medieval boundary and can be seen on the ground and the Ordnance Survey map. The north end of the boundary starts at the footbridge over Lingmell Beck that you probably crossed on your way to Hutching's How. It follows the outer boundary wall of presently enclosed land running roughly south and then south–west all the way to the summit of Illgill Head. It then turned steeply down slope to Wast Water and remnants of a wall are marked on the Ordnance Survey map near High Iron Crag. The western boundary follows the lake shore north-east past Brackenclose and north back to the footbridge. What is now called Wasdale Head Hall farm was probably first built by one of the 17th- or 18th-century lessees of the enclosed area, a date later than other farms at Wasdale Head. Reflecting this history, the farm is named "Fences" in the censuses from 1851 to 1891.

19th-century climbing visitors: Wasdale Head, tucked away close to Cumbria's west coast, was not an easy place to get to in the early 19th century, and its wild and austere landscape was not the mainstream taste of the time. Visitors searching for Pictureque beauty looked elsewhere as did the villa builders who eschewed the lakeside opportunities with the exception of a Halifax merchant who built Wasdale Hall in 1829 at the foot of Wast Water. The visitors to Wasdale Head from the mid-19th century onwards were looking for something completely different: opportunities for hard walking, scrambling and, later, climbing amongst the fells. Local farmers saw the opportunities and embarked on early examples of farm diversification. In 1856, Will Ritson at Row Foot built a wing onto his farm to accommodate visitors and opened an inn called the Huntsman's Inn; later, Thomas and Mary Tyson started taking in guests at the nearby Row Head Farm. Will Ritson was an eccentric and very genial man who fostered a unique Wasdale Head hospitality at his inn until he retired to Down in the Dale. The inn was taken over by Daniel Tyson in1879 and renamed the Wastwater Hotel (now the Wasdale Head Inn). Here and at Row Head Farm the climbing visitors were a mixture of Oxbridge reading parties, alpinists looking to hone their snow- and ice-climbing skills, other upper-middle-class professionals attracted to the climbing scene as well some notable local characters. By the end of the century, rock climbing had become a new sport, a branch of mountaineering, and Wasdale Head can reasonably claim to be the sport's pioneering centre in the UK.

Rock climbers at Wasdale Head in 1893 with the famous pioneer O.G. Jones sitting in the centre (FRCC Archives)

Wasdale Head village green: From where you are sitting you can see the main Wasdale Head car park and, if you parked there before making your way to Hutching's How, you were probably unaware you were parking on a registered village green. The statutory purpose of a village green is to provide space for recreational activities, which does not include car parking. Because the parking is technically illegal, the land has been left unmanaged and other complementary visitor facilities have not been put in place. The result is a mess that probably makes this village green the most defiled in England. This issue is the nub of some difficult visitor-management challenges in the valley. During the 20th century, Wasdale Head broadened its appeal as a destination for forays into the fells for walking, scrambling, fell running and climbing, and then at the end of the century a new phenomenon arrived – large-scale charity ascents of Scafell Pike. There are now some 200,000 annual visitors to the valley bringing many currently unresolved problems for the host community and the environment. New initiatives in visitor management in the valley are under discussion by a whole-valley partnership involving the National Trust, National Park Authority, the local community and others. A Visitor Management Plan prepared in 2013 is awaiting agreement and implementation.

Lambing Knott, Buttermere

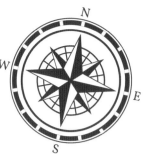

A fine viewpoint overlooking Buttermere, a dramatic valley head and one of the District's most important Herdwick sheep breeding farms; a quintessential Lakeland scene

Location:	NY193155; N54°31.748 W003°14.890
Height:	200 metres
Geology:	Skiddaw Group
Ownership:	Private
Access:	Open access land; either descend from Robinson or make a short but rough ascent from the gate on the road near the homestead
Website panorama:	From south-east through south to north-west

Lambing Knott from the west

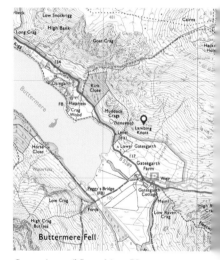

Location of Lambing Knott

⬤ The View

An arc of high volcanic tops from Fleetwith Pike round to Haystacks, High Crag and High Stile creates one of the most dramatic daleheads in the District. Encircled are the flat valley floor of Warnscale Bottom, the inbye pastures of Gatesgarth Farm and the photogenic Buttermere with its shore-side woodlands. This upland valley has a history of relatively low-key settlement and despite its stunning natural beauty has avoided the excesses that tourism can bring in its wake. Although relative remoteness has helped, this is mainly due to benevolent and innovative land acquisition and management in the 19th and 20th centuries, which have kept the developers at bay and retained its scenic beauty.

View from Lambing Knott looking west

Your viewing position: Looking west towards Burtness Comb

Geology and Burtness Comb: Within this view, three of the Lake District's main rock groups meet and display different terrain that can be identified fairly readily. The highest summits in front of you, from Fleetwith Pike round to High Stile, down to the base of their higher crags (roughly the upper third) are rough, volcanic rocks of the Borrowdale Volcanic Group. The lower slopes of those tops and the side of the valley where you are sitting are noticeably less craggy; these are composed of rocks of the older Skiddaw Group. To your right on the slopes above Burtness Wood, including Dodd, cragginess is replaced by smoother profiles marking the start of the Ennerdale granite intrusion that outcrops north-east round to Scale Force and then over into Ennerdale.

During the Ice Age, the Buttermere valley was filled with a large glacier moving left to right and gouging out the deep valley trough. Opposite you, in Burtness Comb, a short tributary glacier eroded its valley floor much more slowly and would probably have joined the main glacier as heavily crevassed ice falls. With the ice gone, this small tributary glacier can be seen to have carved out a spectacular bowl-shaped combe ringed by steep cliffs with Comb Beck falling steeply in its hanging valley down into Buttermere.

Buttermere: Buttermere and its twin lake, Crummock Water, are in separate over-deepened rock basins gouged out by the glaciers that descended Gatesgarthdale Beck and Warnscale Beck from the high central peaks to combine and create one of the Lake District's radial glaciated valleys. Buttermere is quite deep for its size, at just over 28 metres maximum. At the head of the lake there has been substantial deposition of post-glacial alluvium, suggesting the lake was probably larger originally. From where you are sitting you can see two good examples of fan deltas on the north and south shores where the waters from the hanging valleys of Hassnesshow Beck and Comb Beck enter the lake. The marked difference in size is largely due to the different character of the rock particles that make up the deposited material. Comb Beck, despite a slightly larger catchment and steeper profile, has produced a much smaller delta because the eroded hard volcanic rocks have yielded coarser, more angular particles that transport less easily than the more finely eroded, slatey Skiddaw Group rock brought down by Hassnesshow Beck. Buttermere, which is owned by the National Trust, is a protected Site of Special Scientific Interest on account of the lake's vegetation, fish and other fauna. It is considered to be one of the best examples in the Lake District of a healthy, low-nutrient-level lake. This has come about because human activities have had little adverse impact within the lake's catchment; it is an example of the good condition displayed by all the lakes in the District until relatively recently.

Romano-British farmstead: Immediately down to your right, close to the road, a "homestead" is marked on the Ordnance Survey map. This is the remains of a Romano-British farmstead that is protected as a scheduled ancient monument. It is the only significant evidence of early human occupation in this upper part of the Buttermere valley. This is an example of a small, non-defensive, enclosed farmstead found in many locations in Cumbria. It is of a form that pre-dates Roman arrival and includes a sub-circular enclosure containing two hut circles.

Gatesgarth: The land of Gatesgarth farm below and around you has a documented history from the mid-13th century when it was a vaccary within the Forest of Derwentfells. By this time the use of the more remote uplands as hunting forest was declining and interest was turning to using the land for stock rearing. Although sheep rearing was the more usual option, Gatesgarth was developed as a cattle farm, one part of a Cockermouth-based estate that included other land used for arable farming and sheep rearing. Evidence of the Gatesgarth vaccary's buildings was found during archaeological investigations in 2007–2008: the remains of a longhouse, a trackway and a timber building. The very large encircling walled enclosure (within which you are sitting and clear on the Ordnance Survey map) that extends from the valley floor, part way up Gategarthdale Beck, up to the ridgeline at Robinson Crag and then down Hassnesshow Beck is thought to be the Gatesgarthside referred to in medieval documents. Stock enclosures referred to as a "side" are found elsewhere in association with vaccaries, such as Thin Side in Patterdale (see the pages on Bell Knott, Patterdale). From the late-medieval period, the land was let to a succession of tenants and the land around the head and shores of the lake was enclosed. By the middle of the 19th century the land was a Herdwick sheep farm solely occupied by the Nelson family who from 1850 to 1934 improved both the land and the stock to gain a Lake District-wide reputation for Herdwick breeding.

The Buttermere Round: In the latter decades of the 18th century, the first tourists were arriving in the Lakes, many staying in Keswick. One of the more adventurous tours arranged from the hotels was the Buttermere Round, which passed below where you are sitting. It was a day-long trip to Buttermere either by way of Newlands or through Borrowdale and over the Honister Pass. On the latter circuit, after passing awestruck through the Jaws of Borrowdale, the ascent of the Honister Pass would have been an epic adventure in a horse-drawn coach up a steep, winding track, which necessitated the passengers getting out to walk and even push the coach at times. The descent was equally fearsome, controlling the coach and horses while passing below the intimidating, dark precipice of Honister Crag. To the relief of all, the coach would eventually arrive at Gatesgarth and the tranquil waters of Buttermere; then on to the Fish Inn in the village for lunch.

Coaches descending the Honister Pass

Enhancing beauty: Although during the latter part of the 18th century the beauty of Buttermere and its surroundings was becoming appreciated, it appears to have been a landscape with little in the way of woodlands. This then began to change. Just before 1800 Hassness House, the only villa beside Buttermere, was built by Thomas Benson, a Cockermouth lawyer, with gardens of exotic species and some woodland, mainly larch (the present house is a replacement on a slightly different site). Around the same time Charles Howard, Duke of Norfolk, who owned Gatesgarth Farm and all the land above Buttermere in front of you, established a plantation of Scots pine on the land at the head of the lake, much photographed today. (The Duke of Norfolk was equally active on the shores of Ullswater – see the pages on Bell Knott, Ullswater.) Then in 1815 John Marshall, a Leeds industrialist, acquired Gatesgarth and other properties around Buttermere (and also around Crummock Water and Loweswater) and, with the support of William Wordsworth, set about planting new woodland to enhance the views and provide some commercial return. By 1845 he had acquired and planted Crag Close adjacent to Hassness to your right, Pinfold Close immediately below Lambing Knott, Toad Pots (the strip below the ascending path to Scarth Gap), Horse Close on the Comb Beck alluvial delta and a large area on the slopes south of Buttermere, marked Old Burtness and Burtness Wood on the Ordnance Survey map.

Conservation covenants: Subsequently in 1934 the successors of John Marshall put much of the landscape in front of you up for sale, setting off a chain of events that brought into prominence one of the National Trust's more potent conservation tools. The impending sale brought together a number of people who were concerned about conserving the greatly loved beauty of Buttermere and its surroundings into what was called the "Buttermere Scheme". Between them, the National Trust, the historian Professor G.M. Trevelyan (chairman of the National Trust Estates Committee at the time), Balliol College, Oxford, and several other private individuals bought all the properties in a co-ordinated series of acquisitions. Subsequently, some of this land was gifted to the National Trust, but other land, notably the huge Gatesgarth Farm that Trevelyan had purchased in July 1935, stayed in private ownership, though subject to restrictive conservation covenants in favour of the Trust. Through this scheme the Trust gained an appreciation of the value of covenants in protecting valuable landscapes anywhere, but the law at the time only allowed it to accept covenants over private land if there was Trust land nearby. Appropriate lobbying in Parliament eventually led to the National Trust Act 1937 that empowered the Trust to enter into covenants on land "restricting the planning, development or use thereof" with any willing landowner; in effect making the Trust a back-stop planning authority.

An interesting curiosity is that the verges (and only the verges) either side of the road between Gatesgarth and the summit of the Honister Pass are in the National Trust's ownership. Evidently, this was a belt and braces acquisition to deter any ill-considered development proposal by preventing access from the road.

Herdwick ewes being brought onto the inbye land at Gatesgarth

The Climbing House: Directly below you is the private property Lower Gatesgarth, built in 1912 by Arthur Pigou, Professor of Political Economy at Cambridge University. He was a keen mountaineer who over many years entertained a host of Cambridge friends and undergraduates with an interest in mountains. Guests included Sir Claude Elliott and Geoffrey Winthrop Young, both presidents of the Alpine Club, Professor G.M. Trevelyan, Wilfrid Noyce, a member of the successful Everest expedition of 1953, and many others. The property's association with a stream of prominent mountaineering visitors led to it being known as The Climbing House. Although a private residence, it takes its place alongside other historically notable establishments, like the Wasdale Head Inn and the Gorphwysfa hotel at Pen y Pass in Snowdonia, that fostered the nascent, largely Oxbridge-focussed, rock-climbing fraternity. On his retirement, Pigou sold the house to Claude Elliott but retained use of it every May, June and July. There is a charming story that in the spring of 1953 the mother, wife and baby of Wilfrid Noyce were staying with Pigou at Lower Gatesgarth while the Everest expedition was underway. Nervously waiting for news, the telephone rang at midnight (presumably the night of 1st–2nd June) with the news that Everest had been climbed. The relief and excitement were enormous and those present reported that Pigou was "with difficulty restrained from giving the baby champagne". The rest of the world heard the news from *The Times* newspaper later in the morning.

Ling How, Bassenthwaite

A good view over Bassenthwaite Lake and the pastoral landscape of Cumbria's north-western coastal lowlands

Location:	NY240300; N54°39.560 W003°10.780
Height:	320 metres
Geology:	Skiddaw Group
Ownership:	Private
Access:	Open access land; a short distance off the path leading to Ullock Pike
Website panorama:	From north through west to south-west

Ling How, centre, from the south-west

Location of Ling How

♀ The View

The view takes in much of Bassenthwaite Lake which, although one of the District's big lakes, is not generally a priority on visitors' itineraries. For many, it lacks the scenic qualities of other lakes but it is nevertheless a popular recreational venue for dinghy sailors and canoeists and its fish attract both fishermen and the stunning visiting ospreys.

The lake docs, however, have a number of pollution issues, which are being addressed by multi-agency actions across the lake's river catchment. The viewpoint also looks out north-westwards over the coastal lowlands beyond the National Park boundary – a pastoral landscape of dispersed farms and hamlets with a long history of human occupation. Across the lake are forests that in their early years did much to damage the reputation of the Forestry Commission but are now managed on very different principles.

View from Ling How looking north-west

Your viewing position: Looking west over the lake

Bassenthwaite Lake: The lake fills a trough gouged out of ancient sedimentary rocks of the Skiddaw Group by ice moving north-west from the central Lake District. At one period the lake was part of a single body of water with Derwent Water until glacial till and river alluvium deposited on a rock bar divided the lake into two. The eastern shores of Bassenthwaite Lake are covered with extensive deposits of glacial till shaped into a distinctive embayed shoreline and a low drumlin at Broadness. Down to your left and marked on the Ordanance Survey map are references to Sandbeds and Sandhill. These indicate a wider locality where stratified sequences of sandy gravels were deposited in a periglacial (near glacial) environment during the short-lived Younger Dryas cold period some 12,000 years ago. Two exposures, a quarry by Sandbed Gill and further south at Throstle Shaw, are geological Sites of Special Scientific Interest because of the key evidence they provide of significant local landscape modification during that period.

Early history: Evidence suggests this landscape has probably been occupied from Neolithic times. To your right, beyond the far side of the lake, is Elva Hill where there is a stone circle probably from the late Neolithic to middle Bronze Age period. Situated in a wide open landscape, it is a scheduled ancient monument comprising a large regular stone circle of 15 pink granite boulders of some 33 metres diameter.

Church of St Bega

In the same direction close to the lake shore is the tree covered Castle How. This feature was developed and used as a hillfort probably during the post-Roman period, the steep ground to north and south supplemented by earthworks on the east and west sides to provide the defences. It is a scheduled ancient monument that in 1995 was reported to be "well preserved", but when inspected again in 2011 was judged to be a monument at medium risk and in a declining condition largely due to damaging vegetation growth. This is an example of a widespread problem. Looking down left close to the shore by Highfield Wood is the Church of St Bega. Named after a Celtic saint, the building dates from about 950 although the site is probably older. It was extensively restored in 1874. There are records of an abandoned medieval village close by.

Settlement and farming: Much of your view is low land on the edge of upland Lakeland, typified by a dispersed settlement pattern, some small hamlets and hedges as field boundaries rather than walls. The field systems you can see are generally ancient and based around a number of former medieval common fields that were generally larger than those in the upland valleys. The former common field closest to you is the land from the edge of Bassenthwaite village westwards to the lake shore. Sheep farming was an important part of the local economy from medieval times onwards and there were several fulling mills in the locality where wool cloth was cleansed, thickened and stretched. Inbye land was gradually enclosed over time and in the 19th century the higher ground of former common land on and around Binsey and Caer Mote was enclosed by Parliamentary Enclosure. By the middle of the 20th century the remaining arable land in Bassenthwaite parish was ploughed for cereal and root crops for animal feed, rather than human consumption, or sown with grass. Nowadays grass is the dominant crop for silage and sheep (and some cattle) rearing, and as a consequence the number of farms and farm workers has fallen dramatically. Redundant farm houses have become homes for non-farming families or holiday accommodation. The remaining farmers have a mixture of owned and rented fields, often some distance apart, and run their farm businesses with an element of public subsidy through environmental stewardship schemes. This change in the way the landscape in front of you is managed is repeated in many parts the District.

Early commercial forestry: The slopes enclosing Bassenthwaite Lake have been a target for commercial forest planting from the late 18th century onwards. Wythop Brows on the west shore, opposite you, was reported to be "clothed to the top in wood" in 1769 and a visitor in 1802 referring to plantations of larch on the east shore praised the efforts "to turn the almost useless hills to advantage and beautify the country by planting them with trees". Further to your left on the far shore, the Manor of Thornthwaite (part of the land where Whinlatter Forest Park is today) came into the hands of Greenwich Hospital in 1735. The Hospital's efforts to plant commercial woodland were initially impeded by difficulties with the customary tenants and it was not until the Thornthwaite Commons Enclosure Award in 1814 that planting really got underway with 1.2 million trees planted on the Comb, Hospital and Beckstones Plantations (as named on the current Ordnance Survey map) over the following two years. In 1832, in the face of criticisms of poor management, the Hospital disposed of its Thornthwaite plantation, together with its Keswick estate, to the enterprising John Marshall of Leeds (see the pages on Brown Knotts, Derwent Water, for more about Greenwich Hospital and John Marshall). Then in 1919 the newly created Forestry Commission, charged with the task of increasing the nation's timber supplies, acquired the land and started its first forestry enterprise in the Lake District; its Thornthwaite estate extended some 20 square kilometres over both sides of the Whinlatter Pass. The regimented planting of spruce and larch created unnatural-looking and inaccessible forests, which brought much criticism and set alarm bells ringing about the harm unchecked state forestry could do to the beauty of the Lake District. The appearance of the Thornthwaite forestry (and also in Ennerdale) was used as evidence that underpinned campaigns in the 1930s and 1940s to prevent state forestry in the central Lake District (see the pages on Castle How, Upper Duddon). In those early days, the Forestry Commission had no formal obligations with respect to visual amenity or public recreation; things are very different today. The Forestry Commission now manages all the forest you can see on the west shore and Dodd Wood on the east with much wider objectives than in the past.

Wythop and Whinlatter Forests: The Forestry Commission essentially has three aims for these forests: to optimise the financial return from timber production, to facilitate a wide range of public recreational activities and to protect and enhance the forests' nature conservation value. In pursuit of these aims are a host of actions too numerous to cover here, but they include restoration of ancient woodland with native broadleaved trees, changes in species composition in response to larch disease, improving the visual appearance of forest edges through careful felling, restocking and regeneration methods, and introducing new recreational and educational opportunities. The visual appearance of Wythop Forest, opposite you across the lake, will change gradually rather than dramatically over coming decades. The forest of mixed conifer and broadleaved species is being managed on continuous cover principles except on the upper levels where there will be some clear felling, but with the felling coupe shapes designed to prevent harsh boundaries. Restocking (with Douglas fir a favoured species) and regeneration of both conifer and broadleaved species will be managed to achieve natural-looking forest edges to improve the visitor experience when walking within the forest and viewing it from outside. Further away to the left of Wythop Forest, is the very popular Whinlatter Forest Park – England's only mountain forest park – which received 237,000 visitors in 2017. The forest environment already provides a host of different recreational experiences that the Forestry Commission says it plans to build on to create a world-class visitor experience befitting a World Heritage Site, including "new diverse and exciting experiences".

Bassenthwaite Lake restoration: The lake may look serene on the day you are visiting, but as with a number of the District's big lakes, the quality of its waters does not match the quality of the view. A major multi-agency Bassenthwaite Lake restoration scheme has been addressing the problems over many years. The lake is designated a National Nature Reserve and a European Special Area of Conservation and is owned by the National Park Authority, which controls boating and fishing through permits; there is no public right of navigation as there is on other big lakes, like Windermere. The lake is an example of a mesotrophic (moderate nutrient status) water body and provides a good habitat for an abundance of water plants. It has extensive undisturbed shorelines with habitat gradation from emergent vegetation in the water to wet grassland, mixed fen, willow and alder woodland well developed around the northern and southern ends of the lake. The extensive flood plain fen at the south end of the lake is one of the best examples in Britain. One of the factors contributing to the lake's important status was that it supported the UK's rarest freshwater fish, the vendace, which became extinct here in 2001, although some stock had been translocated. In a kind of compensation, ospreys returned to the lake in the same year.

Bassenthwaite Lake has the largest catchment of all Lake District lakes, is very shallow with a mean depth of only 5.3 metres and has considerable fluctuations in water levels. These physical characteristics affect the way various kinds of pollution affect the water environment. There are three key issues: water pollution, siltation and invasive species. Water pollution is particularly about phosphorus entering the lake from surface run-off of agricultural fertilisers and from detergents and human and farm waste. The periodic appearance of algal blooms on the water surface is a noticeable symptom. Over recent years some progress has been made in reducing phosphorus levels in the water through grants to farmers to change farming practices, by upgrading sewage treatment works, by identifying poorly managed septic tanks and by encouraging residents and tourists to change the types of detergents used. Other water pollution arises from cadmium, lead and zinc leaching from the spoil heaps of abandoned metal mines within the catchment, like Force Crag Mine at the head of Coledale. Here, the UK's first compost-based treatment scheme for mine waste came into operation in 2014 and is expected over its 25-year life cycle to reduce significantly pollution levels in Coledale Beck and Bassenthwaite Lake.

Force Crag Mine

The problem of siltation arises from sediment washed down from the surrounding fells; sediment accumulation on the lake bed has doubled in the past 100 years adversely affecting aquatic vegetation and fish spawning grounds. Soil erosion has arisen from the removal of tree cover and intensive sheep grazing of the high fells over centuries that has damaged vegetation and soils. The eroded material is washed down slopes but, rather than being deposited on flood plains, the water and sediments are kept moving quickly down man-modified river channels all the way to the calmer waters of the lake. Within the Bassenthwaite Lake catchment, remedial measures are underway or under consideration, such as tree planting, reductions in sheep stocking levels and reinstatement of natural river regimes. The problem of invasive species takes the form both of non–native vegetation, such as Himalayan balsam, and introduced fish species, such as ruff. Various remedial measures are being implemented including bio-security measures by lake users to prevent spread of the invasive species and also by their physical removal.

Loup Knott, Ullswater

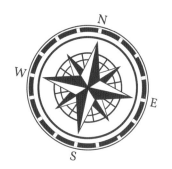

A spectacular airy spot overlooking the northern reach of
Ullswater, its surrounding shores and open landscape

Location:	NY461211; N54°34.951 W002°50.100
Height:	440 metres
Geology:	Borrowdale Volcanic Group
Ownership:	Private
Access:	Open access land; a short steep grassy descent from the crag edge
Website panorama:	From north-east through west to south-west

Loup Knott, skyline left of centre, viewed from the west

Location of Loup Knott

⚲ The View

Loup Knott is part of a line of volcanic crags immediately above pastoral slopes that become
more extensive northwards as the crags lose height and press less closely to the shore. On
the far shore there is a mosaic of enclosed fields and woodland with good soils that in the
past supported much arable farming. This is a landscape that, by virtue of being readily
accessible from the spacious Eden valley to the east, has been inhabited and shaped by

humans from earliest times. In the 18th and 19th centuries, the Picturesque attractions of Ullswater and its surroundings led to the building of numerous villas and ornamental landscapes, most now trading as hotels. Boats have always been an important part of the lake scene, in the past transporting essential goods and people, and today providing leisurely steamer trips and the thrills of dinghy racing.

View from Loup Knott looking north-west

Your viewing position: Looking north-west directly across lake

Rounded fells: Opposite you is a tract of low fells with smooth rounded profiles that are not typical of the Lake District. Great Mell Fell, Little Mell Fell, Soulby Fell and the small Dunmallard Hill at the foot of the lake are the visible remnants of a geological event that tends to get lost in the bigger picture of Lakeland geology. But first, the geological setting: the crag edge where you are sitting and the land for some distance behind you is underlain by hard rocks of the Borrowdale Volcanic Group. Below you, the glacially eroded trough containing Ullswater and the lower land opposite you is underlain by the softer mudstones of the older Skiddaw Group. Far to your right, Heughscar Hill and beyond are part of the residual ring of limestones that surround much of the District. Great Mell Fell and the other distinctive, low, rounded fells opposite are made up of rocks known to geologists as the Mell

Fell Conglomerate Formation. They consist of rounded boulders and pebbles of volcanic origin welded together after being laid down as flood deposits after the volcanic period and before the marine limestones were laid down. Because the conglomerate deposits are fairly uniform in structure, subsequent glacial action has eroded them relatively uniformly leaving the distinctive smooth, rounded profiles.

Early human occupation: It is on the higher ground to your right, rather than on the lake shores, where the main evidence of occupation in prehistory can be found. The land you can see below Heughscar Hill, where a number of ancient pathways cross, is limestone country and the setting for a large number of scheduled ancient monuments. It is not universally agreed exactly what all the remains are or their age, but the evidence points to regular, but sparse, occupation through the Bronze Age to the Roman period. The area named "Moor Divock" hosts numerous Bronze Age remains – the Cockpit stone circle, a stone alignment (a partial double row of upstanding stones) and numerous funerary and clearance cairns. On Askham Fell (Skirsgill Hill) there is evidence of two Romano-British enclosed stone hut circle settlements with associated field systems, and crossing the area is the High Street Roman road. Across the lake and at its foot there are the remains of two Iron Age forts: half

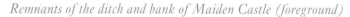

Remnants of the ditch and bank of Maiden Castle (foreground)

right on the rising ground above Rumney's Plantation is Maiden Castle, a univallate hillfort with a roughly circular enclosure, defended by a rampart, ditch and counterscarp bank; and at the outflow from Ullswater is the prominent, tree-covered Dunmallard Hill on top of which is a prominently positioned multivallate hillfort, probably a settlement of high status occupied on a permanent basis.

Settlement and enclosures: The land opposite you on the north shore is the most extensive area of relatively fertile ground in the Ullswater valley and has been sparsely settled from at least medieval times. No nucleated village settlement has developed, although historical documents refer to an abandoned settlement at a location not yet precisely identified. A former common field has been identified immediately south of Bennethead. The absence of a clear community focal point is illustrated by the shifting site for the local church. In sparsely populated rural areas like this, people were served by satellite chapels some distance from the main church town – in this case Greystoke, some five miles to the north. The site of the original early-13th-century chapel is on the lakeshore just north of Skelly Nab where the Old Church hotel now stands. By 1474, the chapel had moved to the site where Gowbarrow Hall now stands and then, after its destruction by the Scots, the "New Kirk" was built in 1558 on the present site, south-east of Little Mell Fell, below Priest's Crag. This building was replaced in 1881 by the present church, which you should be able to make out. From the 16th century onwards, the arable and other inbye land close to Watermillock and northwards between Little Mell Fell and Bennethead Banks was gradually enclosed, often with hedgerows. During the 19th century, Parliamentary Enclosure led to the building of stone walls west of Watermillock (in 1835) and on Little Mell Fell, Bennethead Banks and Soulby Fell. On the south side of the lake, the topography has limited the amount of settlement. Barton Park is the location of a former medieval deer park and the lake-side land is the only land that has been enclosed in subsequent centuries; the open fell where you are siting and behind you has remained unenclosed common land through to the present day.

Designed landscapes: The building of lakeside villas during the 19th century demonstrated an appreciation of Lake District landscapes and provided some social kudos for the wealthy owners. After Windermere, it is on these shores of Ullswater in front of you where there is the greatest concentration of lake-side villas in the District. They were designed to enjoy the views of the lake and surrounding fells from within their own carefully designed ornamental gardens, which usually also included a boathouse. On the north shore, to your left by Skelly Nab, is Hallsteads (now occupied by the Outward Bound School) built in 1815 by the Leeds industrialist John Marshall, whose family was to have a notable influence on Lake District landscapes (see the pages on Brown Knotts and Lambing Knott). Then going round the lake clockwise, other villas are: Leeming House built in the early 19th century, Rampsbeck Lodge built in late 18th century, and Waterfoot. On the south shore by Pooley Bridge, there are Eusemere Hall on an estate acquired in 1795 by the anti-slave campaigner Thomas Clarkson, Waterside House with origins back to 1694, Sharrow built as a fisherman's lodge in 1840 by Anthony Parkin, the son of a former High Sheriff of

Cumberland, and finally Ravencragg, built in the 1840s and the home of Anthony Parkin's brother and nephew, both called William. Most of these villas are now transformed from grand private residences into impressive establishments where today's Lake District visitors can stay in comfort.

Ullswater's water: By the middle of the 20th century, Manchester's water requirements were again proving difficult to satisfy despite the construction of Thirlmere and Haweswater reservoirs. So, in 1961, Manchester Corporation submitted a Private Bill to Parliament to extract substantial volumes of water from Ullswater. This proposal was met with a massive public outcry from local authorities, amenity groups and individuals who raised objections on the grounds of anticipated harmful effects on the environment and recreational activities, the visual impact of changing water levels on the lake's strandline, the lack of a coherent national water policy and inadequate consideration of alternative solutions to Manchester's water problem. In February 1962 an eloquent speech by Lord Birkett, a famous advocate and a past President of the Friends of the Lake District, secured a vote on the Bill's second reading in the House of Lords where it was defeated. Lord Birkett, who died a few days after the vote, became a local hero and had a fell, up above Glencoyne, named after him. However, the vote in the House of Lords did not make Manchester's serious water problems go away. In 1965, the Corporation proposed a more modest scheme to extract water at Gale Bay, down to your right towards the foot of the lake, with an underground pumping station near Parkfoot and a tunnel through to the Haweswater reservoir some distance to the south. Following a public inquiry this proposal was approved but with numerous conditions attached, the most important being to put arrangements in place to prevent excessive pumping. Water cannot be extracted once the lake reaches its lowest natural low-point, thereby avoiding a visually intrusive rim of exposed rock, stones and mud at the strandline.

Boats for transport and pleasure: The wind can funnel strongly along the length of Ullswater, keeping boats tied up or ashore, but assuming you are sitting on Loup Knott in more benign conditions, passing boats will be contributing much to the character of the scene. Boats were an important mode of transport in the Ullswater valley throughout history and remained so until the 20th century. For the resident population, there was a need to move people, general supplies and Royal Mail, and for the local slate- and lead-mining industries there was a need to take the slate and lead ore out of the valley and bring coal, timber, explosives and equipment in. Packhorses, carts and, in the early 20th century, steam wagons using the very poor roads struggled to match the transport capacity of water craft, such as flat-bottomed barges and steam boats. The Ullswater Steam Navigation Company was set up in 1855 to help serve these various transport needs and later evolved into the solely tourist-carrying Ullswater Steamers of today. The company has a fleet of five historic vessels including *Raven*, launched in 1889, which takes its name from Ravencragg, the home of William Parkin who was a director of the Ullswater Steam Navigation Company at the time.

Steam Yacht Raven *arriving at Patterdale*

During the early period of tourism in the Lake District, it was local boatmen who would take the visitors out onto the lake to enjoy the scenery and indulge in amusements, such as the firing of cannons to create impressive echoes off the surrounding fells. By the second half of the 20th century, many visitors were taking advantage of the public right of navigation on the lake and bringing their own craft. With the growth in the popularity of water skiing and jet skiing the conflicts between different lake users, with the environment and the seekers of peace and quiet in the National Park, became hard to manage. In 1975 the National Park Authority banned motorised craft from twenty smaller lakes and tarns and in 1978 secured the approval of bylaws to restrict the speed of motorised craft on Derwent Water, Coniston Water and Ullswater to 10mph. These came into effect on Ullswater in 1983. The result is a view from Loup Knott of sedate and picturesque 'steamers', families in small sailing cruisers and dinghies, kayakers, wind surfers and, if you pick the right day in the summer, the spectacular massed ranks of racing dinghies competing for Ullswater Yacht Club's Lord Birkett Memorial Trophy.

Newtown Knott, Ravenglass

A prominent viewpoint overlooking Ravenglass estuary and harbour, and with views along the West Cumbrian coast

Location:	SD094953; N54°20.730 W003°23.700
Height:	79 metres
Geology:	Eskdale granite
Ownership:	Private
Access:	Open access land; easily approached from the footpath below Newtown Knott
Website panorama:	From south through west to north

Newtown Knott viewed from the north

Location of Newtown Knott

The View

The Lake District is not all high fells and this fine view overlooks the one part of the National Park that touches the open sea. Here, three upland rivers have had their final charge to the sea deflected to force them to combine and create a sandy, dune-protected harbour of some antiquity and nature conservation interest. The landscape has hosted the earliest Mesolithic coastal immigrants, armies defending the Romans' northern frontier, the rise of Ravenglass as a trading port, and its later decline as the silting of the harbour and competition from other ports sapped livelihoods away. It is now a landscape of important coastal habitats, pastoral farmsteads, a small tourist village and quiet harbour, and an incongruous weapons-testing facility.

View from Newtown Knott looking west

Your viewing position: Looking due west across the dunes and sea

Coastline and estuary: Directly in front of you is a fascinating estuarine scene. Prominent are the Eskmeal Dunes to your left and the Drigg Dunes to your right, separated by the Esk channel discharging the combined estuary waters of the rivers Esk, Mite and Irt. This distinctive scene is the result of reworking of thick glacial and other deposits after the end of the Ice Age. These glacial, glaciofluvial and more recent beach and river deposits are everywhere; only at Newtown Knott and the contiguous granite areas north-eastwards is the underlying bedrock at the surface. During the Ice Age this locality was the meeting point of the ice flowing south-west from the high central Lake District and the south-flowing ice from south-west Scotland, which together filled the north Irish Sea basin. At the end of the Ice Age, during the retreat of the glaciers, vast quantities of glacial till and glaciofluvial deposits were laid down onshore and offshore and the three rivers in their over-deepened valleys discharged separately into the sea at some point further west than the current coastline. Over time, the glacial and other sediments were reworked and transported by marine processes, resulting in the development of opposing spits – the Drigg spit developing southwards and the Eskmeals spit northwards. Today, tidal movements make the vicinity of the estuary entrance a zone of coastal sediment convergence. The effect of the growth of the two converging spits has been to divert the course of the Esk sharply northwards and the Irt and the Mite southwards. More recently, in the last 1,000 years or so, a period of deposition of wind-blown sand led to the creation of the Drigg and Eskmeals dunes on top of the two spits.

Coastal habitats: The three-river estuary waters and shoreline and the dunes atop the Eskmeals and Drigg spits are part of the designated Drigg Coast Site of Special Scientific Interest and Special Area of Conservation. The citations refer to the very broad range of maritime habitats with fixed and mobile dunes, dune slacks, intertidal mud and sand flats, vegetated shingle and large areas of dune heath and saltmarsh. These habitats support a very large number of specialised plants, insects and amphibians. The Drigg dunes used to host the largest colony of nesting black-headed gulls in Europe in the early 1980s, but these have now gone. Some suggest that discharges from the nuclear sites to the north may be to blame, while the official judgement is that in 1984 the impact of uncontrolled foxes, myxomatosis amongst rabbits and a record dry spell from May to July combined to prompt desertion.

The condition of Sites of Special Scientific Interest is assessed periodically by Natural England. In the most recent survey (in 2011) the estuary was deemed to be in a "favourable" condition, but the Eskmeals and Drigg Dunes were labelled "unfavourable" with an area adjacent to the Eskmeals Range actually judged to be "declining". One of the problems is the under-grazing of the dunes over many years, which has led to over-stabilisation (some dunes are naturally mobile) and rank grasses dominating dune grasslands and slacks. Attempts to address these issues are guided by a Site Improvement Plan coordinated by Natural England in partnership with landowners and others.

Early settlement history: To your left along the coast just south of Eskmeals, there is evidence of occupation by Mesolithic peoples who probably lived on fish, shellfish and wildfowl to supplement food gathering and hunting. But a resource of particular interest appears to have been flint, with substantial quantities of "microliths" (very small flint blades, scrapers and arrow heads) found at several sites, including Monk Moors and Williamson's Moss. The flint appears to have been carried ashore from an uncertain source, possibly a chalk reef under the Irish Sea. Subsequently, Neolithic and Bronze Age peoples occupied the coastline at a slightly lower level for essentially the same natural resources, but probably, like the Mesolithic peoples, on a seasonal rather than a permanent basis. In the vicinity of Brighouse Farm, immediately down in front of you, geophysical surveys have identified possible Iron Age or Roman features and further right behind the bank of trees is the site of the Roman fort Glannaventa and the adjacent "vicus" (civilian settlement).

The former beacon on Newtown Knott

Newtown Knott.   Ravenglass.

The Romans arrived at Ravenglass in 120 AD and the fort appears to have been built on the site of an earlier fortlet – part of a line of such fortlets extending coastal defences southwards from Hadrian's Wall. Ravenglass is part of the UNESCO Frontiers of the Roman Empire World Heritage Site. The particular attraction for the Romans was the harbour for importing and exporting goods, some of which would have been carried on the road up Eskdale to the fort at Hardknott, over the pass and on to their fort at Ambleside.

Ravenglass harbour and beacon: To your right you can make out the buildings of Ravenglass. In 1208 the village was granted a market charter and gradually grew to become a busy market centre and prosperous port whose fortunes lasted until the 19th century. In Elizabethan times the harbour handled general cargo and fishing boats and during the 16th and 17th centuries it was important as the only natural harbour between the rivers Dee and Solway. Loading and unloading was done at wooden jetties and mooring posts, some on the shore at Ravenglass and others in deeper water opposite you on the far shore of Eskmeals Dunes. Trade was in slate, iron ore, coal, grain, cattle, wood and various other products and included significant international trade, such as corn from North America. There was also a certain amount of illegal trade in brandy, rum, sugar and tobacco via the Isle of Man tax haven. Then by the late 18th century there are records of ships' captains failing to pay harbour dues, suggesting the port was getting into difficulties.

In 1823 efforts were made to reinvigorate trade by making entering and leaving the harbour more attractive to mariners. A beacon tower was constructed on Newtown Knott to act as a leading mark and new sailing directions were published. The sailing directions advised ships' masters to cross the bar just off the estuary entrance by bringing the Newtown Knott tower and a beacon on the Eskmeals dunes into line, then to proceed into the estuary and drop anchor when the tower was on a compass bearing south-east by east and finally to moor to one of the mooring posts on the Eskmeal dunes. Ships were loaded and unloaded at low tide from carts on the sands. Despite the efforts made and a period during the 19th century when local slate, granite, copper and iron ores were brought to the harbour by rail, it never regained its former importance; the combination of silting of the harbour and long-standing competition from the port at Whitehaven just up the coast undermined its trade. The last ocean-going vessel to bring a cargo into the harbour was the schooner *Isabella* in 1914, which was laden with fertiliser (guano) from South America. The tower was demolished in the 1940s and the remnants remain on Newtown Knott's summit.

THE LAST SHIP TO VISIT RAVENGLASS 1914   Silverpoint Copyright

The schooner Isabella on
her last visit in 1914

Muncaster Castle, park and gardens: Where you are sitting is just within the Historic England Grade II* listed park and gardens of Muncaster Castle – a grand house and designed landscape in a setting very different from the mainstream of Lake District villas and their ornamental gardens. A castle was built on the site in the late 13th century and then extended in the 14th century when a pele tower was built on Roman remains. It has been the seat of the Pennington family since 1208 and was largely rebuilt in the mid-19th century. The gardens have 18th-century origins with terraces and pathways designed to take advantage of views of the Esk valley and fells. The grounds contain many exotic species and there are extensive areas of planted woodlands including below to your left and further left on the north bank of the Esk. The deer park, which extends almost as far as Newtown Knott, has an almost complete enclosing wall and is referred to in documents as far back as the 1500s.

Remote installations: To your left, south of the Eskmeals dunes, is an area of woodland interspersed with buildings adjacent to the shore. This is a Ministry of Defence weapons-testing facility that was first established as a heavy-gun range way back in 1897. It is now operated by a private company on a 25-year government contract. During the 20th century, successive governments viewed the West Cumbrian coast as a remote area where facilities best kept away from centres of population could be located. Along the coast to your right, a little beyond your view at Drigg, is the nation's Low Level Radioactive Waste Repository and further north again is the site where the UK's first commercial nuclear power station was built in the 1950s, now the location of the massive Sellafield nuclear-waste storage and former reprocessing plant.

Future sea-level changes: This landscape has been highly mobile since the Ice Age and will doubtless continue to be so, raising questions about the potential impacts of the anticipated rise in sea levels in the medium- and long-term futures. Such matters are currently being looked at during the preparation of a new Cumbria Coastal Strategy that is being coordinated by Cumbria County Council. In the interim, until the new strategy is completed, the Cumbria coast is covered by Shoreline Management Plans that aim to manage the risks of coastal flooding and erosion and to identify the areas where measures to protect the coastline represent the best value for money. The shorelines of the coast and the estuary in front of you have been assessed and one of three policies applied – "Hold the Line", "Managed Realigment" or "No Active Intervention". The policy of "Hold the Line" is only in place at the village of Ravenglass (built on reclaimed saltmarsh) where flooding has taken place in the recent past and physical defences are in place. The policy of "Managed Realignment" applies to the shoreline around the Eskmeals Dunes and Range, and for all the rest of the coast and estuary shoreline within your view the policy is "No Active Intervention". Under these policies there is recognition of an increasing risk of flooding of some properties over the next 20 years, and, looking 100 years ahead, some property abandonment due to flooding or inaccessibility. Additionally, prehistoric, Roman and later historic sites are likely to be at risk – for example, the western side of Ravenglass Roman fort, which has suffered erosion from coastal processes over recent decades.

Pike Howe, Langdale

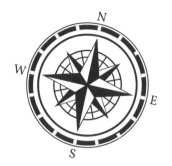

A dramatic rocky promontory beneath the rugged volcanic heights of the Langdale Pikes overlooking a fine glaciated valley and a classic upland farming landscape

Location:	NY289068; N54°27.118 W003°05.933
Height:	400 metres
Geology:	Borrowdale Volcanic Group
Ownership:	Private
Access:	Open access land; readily accessible from a footpath on the north side of Dungeon Ghyll
Website panorama:	From north-east through south to south-west

Pike Howe with Pike O'Blisco beyond

Location of Pike Howe

The View

Before you is one of the finest examples of a traditionally farmed upland Lakeland valley. Dominated by the distinctive Langdale Pikes, it is a steep-sided, twisting, glaciated valley where small piles of rock chippings, field boundaries, farm buildings, cottages and ancient conifers provide clues to the history of human occupation from Neolithic times to the National Trust tenant farmers of today. The legacy is a classic view of a cultural landscape fashioned by Lakeland's agro-pastoral upland farming traditions that is now safeguarded thanks to past initiatives to resist development proposals and acquire farms for transfer to the care of the National Trust.

View from Pike Howe looking east

Your viewing position: Looking south across the valley towards Side Pike

Volcanic Langdale Pikes: The startling profiles of the Langdale Pikes, when viewed from lower down the valley and from Pike Howe, suggest more obviously than anywhere else in the District that this is volcano country. What you are looking at here are the remains of part of a major volcanic centre that collapsed in on itself after its active period to form what geologists call a "caldera" – in this case the Scafell Caldera, which was a depression, largely filled with a lake, taking in the upper reaches of Langdale, Eskdale, Wasdale and Borrowdale. The particular part of it where you are, encompassing the Langdale Pikes and Pavey Ark, is designated a geological Site of Special Scientific Interest because it displays, according to the citation, much of the sequence of extrusive volcanics and, above these, "a superb caldera lake sediment sequence of volcanically derived sandstones, siltstones, conglomerates, breccias and tuffs". The reason this location is so special from a geological point of view is that the preservation of subaerial (that is, on the land surface) volcanoes in the geological record is rare. In most cases, the rocks would either have been eroded away or remained buried; but here, unusually, the detailed structure of a subaerial caldera volcano is readily accessible for geologists to study.

The twisting valley: You can see from where you are that the valley of Great Langdale displays marked changes in direction: it is not a straight glacial trough. This is relatively rare in Lake District valleys and is a good example of a location where a series of faults has been the main determinant of the valley alignment. In Mickleden, round to your right out of sight, the valley alignment is south-east following the line of the major Rossett Gill Fault that runs from Great Gable right through to Little Langdale. Where the valley floor comes into your view and runs below you, the main glacier has turned east-north-east on the same alignment as Oxendale to follow the main structural west-south-west to east-north-east anticlinal axis in the Borrowdale Volcanic Group rocks. The valley then turns south-east again, directly away from you, to follow a north-west to south-east fault, which is also exploited by Dungeon Ghyll.

Neolithic stone working: Down to your left, adjacent to the road just beyond Harry Place farm, are the Copt Howe boulders on which there are two panels of rock carvings thought to be of late-Neolithic date. Above you are the sources of stone for Langdale's famous Neolithic axes that were produced between 5,800 and 5,300 years ago and have been found at many sites across the country. It is not too fanciful to imagine Neolithic man following a track up out of the valley past Pike Howe to reach this rare resource from which were fashioned beautiful, sharp stone tools that did much to change the originally forested landscape. The commonly used reference to a Langdale "stone axe factory" is a little misleading because the axes were not fully fashioned here, nor was there a single working location. The source rock is a fine-grained volcanic tuff that, when struck, chips conchoidally (shell-like) enabling it to be easily shaped. It lies in a continuous, sinuous outcrop between the 500 and 900 metre contours running from Stickle Tarn, through the Langdale Pikes, Bowfell, Scafell Pike and northwards to Glaramara. Along this outcrop, scores of working sites have been identified where the rock was hewn and crudely shaped as rough-outs for later shaping and polishing into axes and adzes at lowland sites some distance away.

Ring garth and enclosure walls: There is little evidence available to allow a description of the valley's history from the Neolithic to the medieval period. However, thanks to research undertaken by the National Trust, we have a fair understanding of medieval and later settlement in the valley, particularly through evidence derived from field boundary walls and farmsteads. A document of 1216 refers to the "inclosed land of Great Langden", which is indicative of the existence of a ring garth that separated the tenanted farmland on the valley floor, cultivated in strips as an open field, from the stock pastures on the manorial waste on the

Remnants of the medieval ring garth at Kirk Howe

open fellside. The probable line of this ring garth has been traced on the ground. Side House, opposite you across Great Langdale Beck, lies just inside the identified line of the ring garth that comes in from the left and then follows the line of the present-day permitted footpath rightwards. It continues at the same general height to cross the Blea Tarn road just above Wall End Farm, and is then out of sight to Oxendale Beck, above Stool End Farm. On your side of the valley the ring garth has been identified following the line of the present road from Ellers up the valley towards you, then diverging from the road to Millbeck Farm, then running just above Rossett and immediately below the mound of Kirk Howe (which lies in a direct line with Rossett Bridge from where you are sitting) and on to Middle Fell Farm. On the ground the surviving ring garth can be seen as footings and in places as a standing wall.

In the 16th century, the rising population in the valley encouraged the enclosure of irregular walled intakes of land for food crops, particularly on the outside edge of the north side of the ring garth – for example, the fields below you rightwards from the Sticklebarn car park between the described line of the ring garth and the track followed by the Cumbria Way. The remnants of a wall on the left side of Kirk Howe and another further left are possibly from this period. Then in the 17th and early 18th centuries there was major intaking of fellside above the 16th century intakes and elsewhere immediately outside the ring garth, by entrepreneurial yeoman farmers (the so-called "statesmen"), mainly for cattle grazing rather than growing crops. They enclosed what was technically common grazing land, but in practice had already been shared out between them for individual use. Over the same period there was the gradual enclosure of the common cultivated land on the valley floor within the ring garth for the sole use of specific farms. By the middle of the 18th century, much of the valley floor below you and away left downstream was enclosed and ceased to be common field, although the line of a valley through route was retained. A section of this route is now the prominent unclassified road on the valley floor coming towards you. (The present main valley road, based on a much older road that linked the farmsteads, was built in the 1930s.)

Then by the end of the 18th century, major changes in the rural economy had undermined the livelihoods of the yeoman farmers and many of them were forced to sell out to other more successful landowners. Consequent land-management changes included substantial embanking works to contain Great Langdale Beck and further land enclosures implemented by mutual agreements and Act of Parliament. The last remaining areas of common field (mainly the land from Rossett Bridge rightwards, including the Langdale campsite) were enclosed, like other parts of the valley floor, mainly by hedges rather than walls. The fellside opposite, on the north flanks of Lingmoor and Side Pike, were also enclosed with boundaries of straight walls and right-angled corners typical of the time. All the fellside opposite you is no longer common land, while on your side of the valley the enclosed land does not ascend so far and the land above it remains common land today (Great Langdale Common).

Farms past and present: One of the notable features of the valley landscape you are looking at is the scatter of individual farmsteads and dwellings on the valley flanks. Although seven farmsteads have been abandoned since the period of the yeomen farmers, the remaining farms and former farm buildings tell the valley's story in much the same way that the field walls do. Immediately below you, just west of the National Trust car park, in the trees, is Rossett, which may have been the site of a Norse shieling that was later permanently settled in the medieval period. Across the valley floor is Side House, which is thought to have medieval origins. Then looking left, down valley, there is a noticeable curving line of dwellings on the north side of the valley, particularly striking when the sun catches their white walls. These buildings have origins as medieval and Tudor farmsteads facing south with a readily available fellside water supply and above the wetter parts of the valley floor, but with ready access to the inbye land just below. Amongst the farmsteads along this northern side of the valley, the name "place" is thought to indicate a medieval farmstead that was a settled encroachment (known as an "assart" and often named after the coloniser) immediately outside the ring garth. Examples include Robinson Place, Harry Place, the former Johnson Place (in the vicinity of Raw Head and Pye Howe today) and Middlefell Place (now Middle Fell Farm). Several of these farms have become the core of the National Trust's large estate in Langdale.

National Trust farms: For the first three decades of land acquisition by the National Trust in the Lake District, it was largely land of special scenic merit that came into its care. However, in 1929 the gift of Wall End Farm, Stool End Farm and the Old Dungeon Ghyll Hotel (a former farm) was accepted by the Trust signalling that, beyond scenic beauty, it was committed to conserving the Lake District's farming families' livelihoods, their Herdwick flocks and all the associated traditions. The person who provided these first Langdale acquisitions for the Trust was Dr G.M. Trevelyan, a successful author and professor of history at Cambridge University at the time. He had come to love the valley after acquiring Robin Ghyll, a 17th-century farmhouse near Chapel Stile, as a family holiday home just before the First World War. He bought the two farms and the hotel and gifted them to the Trust in order to protect the character of the upper reaches of Langdale, which a few years before had been threatened by a proposal for a holiday-hut encampment far up in Mickleden. Since then other farms have come into the National Trust's ownership, such as Millbeck Farm and Harry Place in 1944 (the latter another gift from Trevelyan), Side House in 1963 and Robinson Place in 1974. The Trust now owns much of the valley below the unenclosed fell land. What you have in front of you is an upland farming landscape that serves as a showcase of how the distinctive agro-pastoral traditions of the Lake District have created scenes of very special character that have been deemed worthy of World Heritage status.

Ploughing at Millbeck Farm with Pike Howe above; early/mid 20th century

Juniper and gunpowder: Opposite you on the flanks of Lingmoor Fell, above Side House farm, are extensive stands of juniper interspersed with a few trees. This juniper has an interesting link with the Langdales timeshare development further down the valley, built on the site of the former Elterwater Gunpowder Works (in the trees to the left of the active Elterwater slate quarry). Gunpowder was made from saltpetre and sulphur mixed with charcoal. The saltpetre and sulphur were imported from abroad through the port of Greenodd on the south Cumbria coast and transported up the full length of Windermere. The charcoal came from local woods, with juniper a particularly valued resource that was quite possibly planted specifically to meet this demand. The gunpowder works, which served a market in the local slate quarries and copper mines, closed in 1928 and, following decontamination, was later redeveloped as the present timeshare development.

Rascal How, Coniston

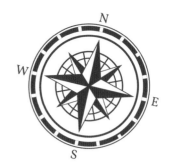

An excellent vantage point looking out over Coniston village and along the length of Coniston Water

Location:	SD 297984; N54°22.542 W003°05.018
Height:	310 metres
Geology:	Borrowdale Volcanic Group
Ownership:	Private
Access:	Open access land; approach by walking south-eastwards from the public footpath that leads up to Hole Rake or by climbing the Mouldry Bank/Rascal How grade-two scramble
Website panorama:	From south-east through south to north-west

Rascal How with Coniston Old Man beyond

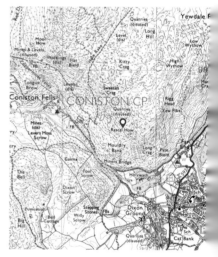

Location of Rascal How

⚲ The View

On the very edge of craggy volcanic fells, this view looks out over a softer, low-lying landscape with a narrow glacial trough cradling Coniston Water. Close by are the raw remnants of past mining activity in the Coppermines Valley and immediately below is the village of Coniston, which expanded to house the miners and their families. Across the lake were the much-exploited medieval lands of Furness Abbey that, from the late 18th century onwards, were covered with commercial forests and are now managed as the Forestry Commission's Grizedale Forest Park. Coniston Water, for a while a venue for fast power boats, is now a recreational centre for boating of the gentler kind.

View from Rascal How looking south

Your viewing position: Looking south-east across the lake

A geological edge: You are sitting on a precipitous edge that marks the boundary between the high fells of the Borrowdale Volcanic Group rocks and the low sedimentary Silurian rocks of the Windermere Supergroup. If you look to your right you can see this edge quite clearly going away from you towards the south-west, reflecting the north-east/south-west structural orientation of folding created by the Caledonian Orogeny. Whilst looking along this edge, pick out in the distance a small but sharp, pyramid-shaped summit, High Pike Haw, and closer to hand in the same direction, Timley Knott. These features are part of a narrow band of rocks, in the past referred to as the Coniston Limestone, that lie between the main sequences of Borrowdale Volcanics and Windermere Supergroup rocks. This band of rocks acts as a marker of this geological boundary and runs right across the District from Broughton Mills in the south-west to Shap Wells in the north-east. The limestone within this sequence gained prominence in the past as a local resource for liming enclosed land

Timley Knott showing the rock sequence formerly known as Coniston Limestone

and for mortar, but geologists now refer to these rocks as the Dent Group because most of the sequence is not in fact limestone, but mudstones, calcareous siltstones, sandstones and rhyolitic tuff.

A prehistoric landscape: If you look again in the direction of High Pike Haw, there is a large tract of land from Banishead (below Timley Knott) through to Bleaberry Haws that is full of evidence of prehistoric occupation. This broad, beck-dissected bench is at around 250 metres and is one of several examples of sites that came to be occupied at around this height away from the coastal zones during the Bronze Age. The extensive remains, mostly scheduled ancient monuments, include burial cairns, small stone circles, enclosures and evidence of former fields. Lying across Bleaberry Haws is a linear dyke, part of a pasture and grazing enclosure, and at Banishead in 1909 there was a find of burial urns with bone ash and a piece of woollen cloth. Some of these finds can be seen in the Ruskin Museum in Coniston.

Medieval settlement: Staying on the west side of the lake, directly in front of you down on the shore by Lands Point is Coniston Hall, a key link to the medieval period. The first Coniston Hall was built by Richard le Fleming after he became the lord of the manor of Coniston in 1250. A deer park was established immediately south of the house taking in what is now named Bleathwaite Pasture, Park Coppice and Bleathwaite Coppice. The present Hall, owned by the National Trust, was built around 1580 and is one of the best surviving examples of a manor house in the Lake District. Settlement across the medieval manor is reflected in single ancient farms and small irregular fields that survive down below you around Coniston. There is no evidence of a ring garth, but a former open field had been established by the late 13th century between Coniston and Coniston Hall where arable fields were farmed in common. Over later centuries there was gradual intaking of land up the lower slopes of Coniston Old Man with the land parcels referred to with distinctive names – "tranearth" extending up the fell north-west of Torver and "scrow", directly across Church Beck from where you are sitting (both marked on the Ordnance Survey map).

Furness Abbey lands: The medieval period on the far side of the lake had a different context. All the land of the Furness Fells between Coniston Water and Windermere was acquired by the monks of Furness Abbey in 1127. They began by developing their interests in the land for hunting and pasturage for sheep and cattle, and established a network of farms across High Furness. A manor farm was established at Hawkshead that became the centre of the Abbey's operations in the area as involvement in the wool trade increased in scale and reach, even into Europe. By 1292 the Abbey was developing interests in the iron industry as well, leading to the establishment of bloomeries using charcoal from the local woodlands. Woodland was being devoured on several fronts: to create sheep and cattle pasture, by the growing numbers of tenants (many of them squatters) exercising their rights to wood for building and daily needs, and to provide charcoal. In 1339 the Abbey obtained the right to make enclosures, which led to the creation of a number of "parks" across High Furness, such as at Lawson Park across the lake opposite you (up above Brantwood) and also Parkamoor and Water Park further south. Within the enclosed lands, arable farming and coppice woodland management could take place without interference from pastoral farming activities and nibbling sheep, and iron bloomeries are also thought to have been established. Following the dissolution of Furness Abbey in 1537, the parks and the sheep farms were let and sold off over time by the Crown. In secular hands, woodland exploitation and iron processing continued with the destruction of woodland so severe that in 1567 a Royal decree placed a temporary local ban on iron processing. Recent surveys have identified the remnants of 160 charcoal burning platforms on the steep wooded land falling down to Coniston Water's east shore. Woodland industries sustained their economic importance in High Furness over the following centuries and then in 1937 the Forestry Commission acquired much of the wooded land opposite you.

Coppermines Valley: To your right is a stark but fascinating area of past industrial activity on which Coniston was founded. Here the Borrowdale Volcanic Group rocks were exploited during the late 18th and 19th centuries for slate that helped roof the nation and copper that gave speed and longevity to the hulls of the British navy. Up on the north-east flanks of Coniston Old Man are the decaying reminders of mining for green slate that was won largely from underground workings, some vast in scale, before being taken down to jetties on Coniston Water for export. An opencast slate quarry is still working at Low Brandy Crag below Crowberry Haws. On the floor of the valley you can see a large tip. This is a conspicuous marker of past copper mining: waste rock dumped after copper ore has been extracted. Veins of copper lay in the area from the head of the Coppermines Valley (just out of sight from where you are, but below the conspicuous Kennel Crag) across to Levers Water. Initially the veins were worked in late 16th and early 17th centuries by the German miners of the Company of Mines Royal (read more in the pages on Brown Knotts), but the period of greatest activity and prosperity was between 1830 and 1860 when Cornish miners were part of the workforce. Most of what can be seen of past mining activity comes from this period – levels, shafts, waterwheel pits, tramways, buildings and waste tips. At its peak, some 600 men, women and children were employed producing 250 tons of copper a month. Copper is difficult to smelt so the ore was exported, at first to St Helens and then later to the UK's main smelting centre at Swansea. The ore was carried down to a lakeside quay by horse and cart (where the Boating Centre is now), taken by boat to Nibthwaite at the foot of the lake, then by horse and cart to the harbour at Greenodd from where it was shipped south. This arduous export route was made much easier with the opening of the railway to Coniston in 1859. The terminus was directly below you, where the Walna Scar Road and Church Beck enter the village. This was convenient for loading copper and slate, but less so in later years for tourists arriving by passenger train.

Brantwood, home of John Ruskin

Brantwood: Despite its natural beauty and setting, Coniston Water did not attract the villa builders and searchers after the Picturesque in the same way as other lakes. The only two stand-out villas and designed estates on its shores are Monk Coniston at the north end of the lake (not in your view) and Brantwood, directly opposite you above the far shore. Brantwood was bought, apparently unseen, by John Ruskin in 1872 and he stayed there until he died in 1900. Ruskin was an internationally prominent intellectual, art critic and patron as well as a visionary writer and speaker on many social, political and environmental issues. His Lake District surroundings and Brantwood itself stimulated thoughts and ideas about man's relationship with nature that fed into his world view of society, capitalism and the natural environment. Ruskin used the Brantwood estate as a practical test bed for ideas about land management that he constantly drew upon in his writing and speaking. Brantwood is a grade II* listed building and is managed, together with the rest of the estate, by the Brantwood Trust, which maintains and presents it very much as it was in Ruskin's day.

Coniston Water: In the 1950s, the bed of Coniston Water was bought by a private individual who was concerned about future use of the lake; ownership was subsequently conveyed to a trust managed by the parish council. The trust deed requires the lake to be under local control and used for the purposes of public recreation, but does not state what sort of recreation – in particular, what sort of boating. From the early 20th century, the lake, together with Ullswater and Windermere, had been used for power boating, including power-boat speed record attempts. Sir Malcolm Campbell set the world water speed record on Coniston Water in 1939 and subsequently his son, Donald Campbell, raised the record a number of times here before he was tragically killed in his boat *Bluebird K7* in 1967. By 1978, the impact of power boats and water skiing on the lake was controversially judged to be so detrimental to the environment and amenity that a bylaw was introduced imposing a 10mph speed limit. Now the lake is used for low-key recreational boating only, although the tradition of power-boat record attempts continues with the annual Coniston Power Boats Records Week.

Coniston village: Below you is Coniston village, in the past known as Church Coniston to distinguish it from Monk Coniston at the head of the lake. From origins as a community based on livestock farming and woodland industries, the village really developed in the late- 18th and 19th centuries through the expansion of the slate and copper mining industries. The peak population for Church Coniston was 1,324 in 1861. The railway had opened 1859 to serve the needs of the mining industries, but later helped the growth of the village's tourism industry, which is now the mainstay of the village's economy. Looking to the future, the village community has prepared a Neighbourhood Plan, one of the national pilots introduced through the Localism Act 2013, which was formally adopted in 2016 and is now a component of the National Park's Development Plan.

Rasp Howe, Kentmere

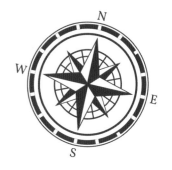

A commanding position overlooking the timeless Kentmere hamlet
with views up into the valley's higher mountainous reaches

Location:	NY467034; N54°25.367 W002°49.384
Height:	370 metres
Geology:	Windermere Supergroup
Ownership:	Private
Access:	Open access land; approach over untracked ground by walking west from the bridleway over Green Quarter Fell; Rasp Howe lies north of the wall
Website panorama:	From south through west to north

View of Rasp Howe, on the skyline, from Kentmere

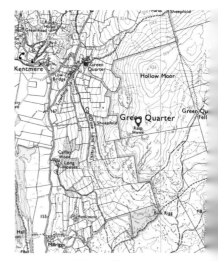

Location of Rasp Howe

The View

This view over Kentmere displays a quiet pastoral scene in a relatively isolated and closed valley. The present-day tranquillity belies a busier medieval past when the valley produced wool, cloth and leather for the nearby trading town of Kendal and was on the main packhorse route from Kendal to Penrith. In more recent times it became a location for mining activity and quarrying slate, which finally came to an end in the mid 20th century. This is now a sheep-farming valley that seems to mind its own business, where the National Trust and the Forestry Commission, big shapers of the landscape elsewhere, are conspicuous by their absence.

View from Rasp Howe looking north-west

Your viewing position: Looking west across the valley

The landforms: Looking to your right up-valley, the boundary between the low relief of the Windermere Supergroup rocks and the craggy rocks of the Borrowdale Volcanics is plain to see. The boundary between the two lies roughly west to east from the Garburn Pass to the pass leading over to Longsleddale. A south-moving glacier has left a legacy of small basins in the valley's longitudinal profile coinciding with rock bars or ridges of moraine. The waterfall Force Jump, just above Kentmere hamlet, marks a volcanic rock bar with the more easily eroded Windermere Supergroup rocks below. After the retreat of the glacier there was a glacial lake immediately above the rock bar that was later filled in with alluvial deposits. Opposite you on the valley floor is Kentmere Tarn, the remnant of a larger lake left impounded in a glacier-eroded basin. Glacial moraines left by the retreating glacier can be made out in several places across the view: to your right immediately north of the track to

Kentmere Hall, on the hillside west of the tarn and to your left around Croft Head.

Early settlement: Within the landscape around you there are a large number of remains that provide tantalising hints of early settlement, possibly from as early as the Late Bronze Age through to the Romano-British period. Three of the sites are protected as scheduled ancient monuments and are marked on the Ordnance Survey map. Down to your left is the Millriggs enclosed hut circle settlement, oval in shape with an enclosure bank up to one metre high within which are the foundations of seven hut circles. Then further up the valley, adjacent to Tongue House, is the site of a similar enclosed hut circle settlement; and further up the valley again, north of the reservoir below Lingmell End, is the site of two small oval enclosures and a number of hut circles. Evidence of early medieval settlement has been found at an archaeological excavation west of the Tongue House settlement in the vicinity of Bryant's Gill. Here a longhouse and other features have been found with a range of radiocarbon dates, including the Norse period, which makes it a particularly interesting discovery.

The medieval Kentmere Quarters: During the Norman period the inbye common lands in the Manor of Kentmere were divided into four hamlets or "quarters" populated with customary tenants who, with the lord of the manor living outside the area, apparently ran most of their own affairs. The names of three of the Quarters still appear on the Ordnance Survey map and the influence of topography on their layout can be seen from where you are sitting. The section of the valley above Force Jump as far as Tongue House provided, within the confines of the steep valley sides, settlement sites and land for some cultivation and the rearing of cattle. Here was Cragg Quarter, west of the river, and Hallow Bank Quarter, east of the river. Then in the length of the valley below Force Jump, the valley sides are less steep and better soils provided the best land in the township. Green Quarter was on the east side of the river and on the west side was Wrea Quarter (later to be referred to as Hall Quarter). Each of the four Quarters had its own land for cultivation and a large enclosed pasture for grazing cattle. In addition tenants from all four Quarters shared rights to graze sheep on the high pastures of the unenclosed Kentmere Common at the head of the valley. Documents from 1710 indicate that historically tenant farmers in each Quarter each held ten "cattlegates" (the right to graze ten cattle on the Quarter's pasture) and grazing rights for 80 sheep on the unenclosed fell at the dalehead. This would have added up to some 5,000 sheep on the dalehead pastures and 150 cattle on each of the four Quarter pastures, suggesting a relatively wealthy pastoral community contributing to the important woollen and leather trades of the nearby medieval market centre of Kendal.

During the following centuries, the area's yeoman farmers held sway, their confidence boosted by a landmark victory in the Court of the Star Chamber in 1623 when local lords unsuccessfully challenged their rights as customary tenants. Land- and stock-management regimes changed slowly until the 19th century. Large parts of the common arable fields on and near the valley floor remained as unenclosed strips until as late as 1834. Parts of Hallow

Bank and Cragg Quarter pastures were divided and enclosed as intake land. Green Quarter pasture remained undivided until the early 19th century, by which time the occupiers of Kentmere Hall had acquired all the rights to the Wrea Quarter pasture at the expense of other properties that had either been transferred to Cragg Quarter or abandoned.

Kentmere Hall: Just to the west of Kentmere hamlet, and clearly separate from it, is Kentmere Hall, a Grade II* listed fortified house that was probably built in the late 14th century by William Gilpin, of the locally prominent Gilpin family. At a later date, a deer park, Kentmere Park, was established across the valley directly opposite you. The house's tower is one of some twenty pele towers in this part of south-east Cumbria that were generally built in response to the threat from raids from the so-called Border Reivers. These English and Scottish family-based raiding groups disrupted life in the unsettled Scottish–English border areas, stealing livestock and other property. There is no firm evidence that Kentmere was ever raided and it is quite possible that the tower was built to convey social status as much as to serve a real defensive purpose. It is now a working sheep farm.

Kentmere Hall

Land improvement: By 1800, the heyday of the yeomen farmers was over and wealthy outsiders started buying up farms and investing in improved agricultural methods. In this valley, the driver for landscape change was solely agricultural improvement and not

Picturesque ideals. There was investment in buildings, land drainage and liming, ploughing and reseeding land. Kentmere Tarn was drained in the 1830s on the, as it turned out, false assumption that good-quality agricultural land could be created. The cumulative effect of all the land-drainage works was rapid, rather than gradual, run-off of water from the land, which seriously interfered with the regularity of flow in the River Kent to the consternation of the many downstream textile and bobbin watermill owners. The Staveley area, including Kentmere, was at this time a significant industrial area based on water power. As a consequence, Kentmere reservoir was built in 1846 in the upper reaches of the valley to regulate the flow of water for the many mills downstream. The watermills are long gone and the reservoir no longer serves its original purpose. The current owners, the papermakers Croppers of Burneside, now maintain it as an aesthetic and community resource.

Another part of the 19th-century improvements was the further enclosure of fell land, notably the Green Quarter pastures within which Rasp Howe is situated. Here, the rectilinear pattern of field boundaries typical of the period is well displayed and contrasts with the seemingly chaotic medieval field boundaries that can be seen adjacent to Kentmere Hall and hamlet. However, as can be seen from the Ordnance Survey map, all the land at the head of the valley beyond Tongue House up to the watersheds remained unenclosed. This is the only common land remaining in the valley today.

Past quarrying and mining: To your left down on the valley floor, a number of factory buildings are visible. Here in the first half of the 20th century was the base of an industry mining and processing diatomite, a white clay created from the fossilised remains of microscopic organisms in the clear water of Kentmere Tarn at the end of the Ice Age. This raw material, when mixed with asbestos and other minerals, created a valuable insulation material. The diatomite was excavated from the land reclaimed from the original tarn in the 19th century and when excavation ceased in the 1980s the new tarn we see today was left behind. In 1955, an oak dug-out boat dated to the early 14th century was found buried in the diatomite by a dragline operator and is now part of the Windermere Jetty collection of boats in the care of Lakeland Arts.

To your left in the vicinity of Millrigg Knott, lead ore was discovered in the mid-18th century. Initially a drift mine was driven south-eastwards far into the fellside on the north-west side of Millrigg Knott and then in the early 19th century vertical shafts were sunk into the same vein from the east flank. The ore was initially taken for processing to a smelt mill at Scroggs, just north of Staveley, but was later taken out of the area by rail. Mining stopped around 1880.

To your right, well up the valley south of the reservoir, are reminders of a cluster of slate quarries that worked good-quality Westmorland green roofing slate (from water-lain bedded volcanic tuffs). Although this slate was quarried from at least the early 18th century, it was not until road communication improved during the 19th century and the railway came to

Retrieving the medieval dug-out boat in 1955

Staveley in 1847 that the industry really developed. There were seven quarries in all. As Welsh slate increasingly dominated the national market, the quarries declined and finally all work stopped in 1956 leaving a legacy of prominent spoil heaps.

River Kent: These now-neglected spoil heaps have brought concerns about pollution of the River Kent, particularly after periods of high rainfall when large quantities of coarse and fine sediment are flushed into the river, potentially affecting both its geomorphology and ecology. The River Kent and its tributaries are designated a Site of Special Scientific Interest and also a Special Area of Conservation indicating the river system's international environmental importance. The SAC citation refers to the river's high water quality with watercourses of rare, protected aquatic plants and endangered, legally protected species, such as the fresh-water pearl mussel and white-clawed crayfish. All are represented in the Kentmere valley. The whole river system has a long-term improvement strategy overseen by a multi-agency partnership that aims to address issues arising from past physical modification of watercourses, habitat decline, diffuse pollution of water, flood risks and invasive species. At the time of writing, Kentmere Tarn is proposed for modification to enable it to function as a River Kent flood-water storage area as part of plans to protect the town of Kendal from flooding.

Silver How, Grasmere

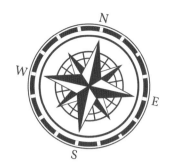

An exceptionally charming view over a valley and village
popularly associated with the poet William Wordsworth

Location:	NY325066; N54°27.047 W003°02.569
Height:	395 metres
Geology:	Borrowdale Volcanic Group
Ownership:	Private
Access:	Open access land; footpaths lead to the summit from all directions
Website panorama:	From north through east to south

Silver How from beside Grasmere

Location of Silver How

Q The View

Before you is a fine example of one of the World Heritage site's scenes of "harmonious
beauty" where clusters of human settlement nestle comfortably alongside lake, river,
woodland and pasture all within the close embrace of modest surrounding fells. To your
right the serenity of the twin lakes of Grasmere and Rydal Water surrounded by woods;

in front, well immersed amongst the trees, is Grasmere village, an international visitor destination with its mix of fine Victorian hotels and residences, cottages, galleries and shops, and then up leftwards towards Dunmail Raise a more open, upland pastoral farming scene, echoing the valley's historic links with the medieval wool trade.

View from Silver How looking north-east

Your viewing position: Looking north-east across Grasmere village

The drainage pattern: The first thing to notice is that the lengths of Grasmere and Rydal Water are not aligned: the river drainage pattern takes a sharp right-angled turn between the two lakes. The explanation for the Lake District's drainage pattern has been the subject of much debate amongst geomorphologists for a long time, although there appears to be general agreement now that it was largely established prior to the Quaternary period – that is, over 2.6 million years ago. There was a belief that the pattern was established in the younger rocks that once covered the District and then superimposed on the underlying older rocks as the younger rocks were eroded away. The current thinking is that weaknesses in rock structures associated with faults and fractures facilitated past erosion by rivers and were thus the main (but not the only) determinants of current drainage patterns. A major fault, the Coniston Fault, runs through the central Lake District along the length of Coniston Water, travels northwards along the valley occupied by the A593, crosses the low col at High Close, goes along the length of the Grasmere valley to Dunmail Raise and then along the length of Thirlmere. This fault has determined the orientation of drainage in the Grasmere valley and Grasmere (lake) itself. However, in the vicinity of Rydal Water there is a west-to-east zone of fractures linking across to a north-to-south fault line running along the Rydal Beck valley that is thought to have had the effect in past geological times of diverting drainage eastwards and then back southwards at the eastern end of Rydal Water.

During the Ice Age, ice flowed down from Dunmail Raise and was joined by glaciers from Greenburn and Easdale to deepen the valley and excavate Grasmere. The south-flowing ice built up and overtopped Loughrigg (creating distinctive lumpy ice-scoured terrain) and joined the Langdale glacier. Against the north flank of Loughrigg, a subsidiary stream of ice was forced eastwards deepening the pre-existing valley and gouging out the basin now filled by Rydal Water.

Early settlement: There is some evidence of prehistoric human presence in the valley. At Allan Bank, the National Trust property on the western outskirts of the village (just hidden by trees from where you are sitting), there is rock art in the form of "cup marks" and the large cairn on Dunmail Raise is also probably prehistoric. (It is commonly held to be the resting place of Dunmail, one of the last Cumbrian kings in the 10th century.) There is evidence of a Norse presence in Old Norse names (for example "Rydal" and "Seat Sandal"), although interestingly many such names do not appear in documents until the 13th century, suggesting they were not introduced into general usage until well after the Norse period. The Norse are likely to have got forest clearance and draining of the valley bottom underway, and by the middle of the medieval period, a pattern of permanent settlement would have been well established. During the medieval period, it is thought that the common field extended from the north shore of Grasmere up the valley towards Town Head with farms scattered along the boundary between the arable and meadow inbye land on the valley floor and the common rough grazing on the fellsides above.

From where you are sitting, you can see quite clearly that Grasmere village is not a single nucleated settlement but a web of interconnected clusters of buildings. This reflects the historical evolutionary growth from individual medieval farmsteads (and possibly Norse shielings before that) into groups of farmsteads and then into hamlets. The main hamlets during the medieval period were Town End (adjacent to Grasmere lake), Above Beck (on the other side of the A591 around Greenhead Gill), Lancrigg (a short way up Easdale to your left) and Town Head (just within your view beyond Helm Crag). Around the parish church of St Oswald was a cluster of farmsteads known as Churchtown that ultimately became the centre of present-day Grasmere village.

Farming reorganisation: By the latter part of the medieval period, sheep rearing and the production of wool was the valley's foremost industry. On the farms, the whole family was involved in washing, carding, spinning and weaving the wool that was then finished at one of at least three water-powered fulling mills that were in the valley at the end of the

15th century. Cloth was taken to the major wool town of Kendal for marketing. By the end of the 16th century the wool trade had declined, and coupled with the excessive subdivision of family land through the generations, the scene was set for considerable reorganisation. Through to the end of the 18th century, the pattern of land tenure changed with the rise of the "statesmen" farmers, new incomer landowners, removal of many tenants, plot amalgamations and land enclosure. The walled enclosures you can see on the valley floor and opposite you below Heron Pike and Stone Arthur have origins largely from this period with some, such as in the vicinity of Town Head, a little earlier. Unusually, what you don't see, though, is evidence of 19th-century Parliamentary Enclosure; you need to look north and east of Ambleside on your Ordnance Survey map to see the familiar large rectangular field boundaries of that period. In the absence of Parliamentary Enclosure, most of the land in your view above the lower enclosed land remains today as unenclosed common land – the Grasmere and Loughrigg Commons.

The early admirers: The poet Thomas Gray visited the Lake District in 1769 on the lookout for Picturesque scenes and wrote in his journal about his descent from Dunmail Raise towards Grasmere where "opens one of the sweetest landscapes that art ever attempted to imitate." Writings like this alerted the educated classes to the intimate, pastoral beauties of Grasmere. By the end of the 18th century, writers, poets and artists were visiting and some admirers of the landscape were building, initially modest, residences as weekend and summer retreats in Grasmere village and along the road to Red Bank below you. Sometime in the 1790s, a Mr Olliff built The Hollens, a villa now occupied by the National Trust, which you can see left of Town End, beyond the A591. In 1799 the poet William Wordsworth moved into a modest cottage in Town End and the first grand villa in the valley, Allan Bank, was built in 1805 by John Crump, a Liverpool merchant on a site that gave a fine view of Grasmere and its island. However, William Wordsworth was not impressed, calling it "a temple of abomination".

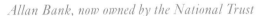

Allan Bank, now owned by the National Trust

Other villas were built later – for example, opposite you on the far side of the A591 is Forest Side built in 1853 for Stephen Heelis, a successful Manchester solicitor. In 1847 the railway came to Windermere, making Grasmere more accessible and stimulating the building of many Victorian hotels and guesthouses. Some villas then started to change hands and were converted into hotels.

Dove Cottage and William Wordsworth: Town End developed as a farming hamlet on the medieval packhorse route between Ambleside and Grasmere that came over White Moss Common. In 1799, William Wordsworth, his brother and the poet Samuel Taylor Coleridge while walking through the hamlet came across an empty cottage that William decided was the place for him and by the end of the year he and his sister Dorothy had moved in as tenants. The garden appears to have been a particular delight for Wordsworth, who described it in a poem as "The loveliest spot that man hath ever found" (words sometimes mistakenly thought to apply to the Grasmere valley). After their marriage, William's wife Mary joined them in 1802 and three of their five children were born there. The cottage had been an inn until a few years before they moved in, referred to as the "Dove and Olive-bough" by Wordsworth. It was not called "Dove Cottage" in Wordsworth's lifetime and the name's first documentary appearance is in the 1851 Census. It was in Dove Cottage that Wordsworth wrote some of his most famous poetry and Dorothy her revealing and much praised Journal. Here Wordsworth wrote one of his greatest poems, "Michael", a heart-wrenching tale set on the fellside across the valley opposite you around Forest Side and Greenhead Gill.

> Upon the forest-side in Grasmere Vale
> There dwelt a shepherd: Michael was his name;

The poem is set in the 1720s or 1730s when the big changes in farming referred to above were underway. It tells a tale of misfortunes for Michael, his wife and their only son and the end of a long family connection with their land that "Was sold, and went into a stranger's hands". Poignantly, after their cottage was swept away, all that was left in the landscape was the remains of a sheepfold that Michael had run out of strength and time to complete.

The idea of conservation: The final and perhaps most important thing to see in the landscape before you is the birthplace of an idea: the idea that landscape should be valued for its own sake and conserved. When UNESCO decided to inscribe the Lake District into the World Heritage List, it adopted a lengthy statement of Outstanding Universal Value that includes the following words:

> The development in the English Lake District of the idea of the universal value of scenic landscape, both in itself and in its capacity to nurture and uplift imagination, creativity and spirit, along with threats to the area, led directly to the development of a conservation movement and the establishment of the National Trust movement, which spread to many countries…

Remains beside Greenhead Gill may inspire thoughts of Michael's sheepfold but are in fact of a 16th-century lead-mining complex

If you had to pick one location in the Lake District that did most to stimulate the new idea of the innate value of landscape, it would probably be here. These scenes not only inspired Wordsworth's poetry, they also shaped his views about landscapes and our relationship with nature: "let Nature be all in all, taking care that everything done by man shall be in the way of being adopted by her". It was the way the newly built Allan Bank and other grand houses brushed against this exhortation that caused Wordsworth to rail against them. Ironically, in 1808 when Wordsworth's growing family needed somewhere more spacious to live, they actually moved into Allan Bank!

Wordsworth first published his thoughts about landscapes in an introduction to a large portfolio of prints in 1810. These writings were subsequently republished in book form in 1822 as a pocket-sized tour guide called *Guide through the District of the Lakes*. However, this volume was much more than a tourist's guide: it was a manifesto about the value of landscapes and how to conserve them. Wordsworth promulgated his ideas through the guide and through his poems (often reprinted in other popular guide books), and then others pursued them energetically over the following decades. One of the most influential during the latter part of the 19th century was Canon Hardwicke Rawnsley, working as a vicar in the Lake District and active in several social and environmental campaigns, who could see that when beautiful places were in the hands of private individuals the risks of harm to them would be ever present. He was very dismayed when he heard in 1893 that the scenic island in the middle of Grasmere, much visited by Wordsworth, had been sold at auction into private hands. This helped propel him into joint action with two other environmental activists, Octavia Hill and Sir Robert Hunter, to found the National Trust in 1895. The most recent owner of the island, on hearing about this history, bequeathed it to the National Trust in 2016.

Thorny Knott, Swindale

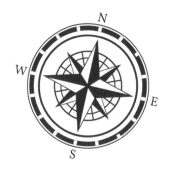

A good valley-edge position where a seemingly 19th-century rural view of Swindale is now the setting for some startlingly innovative 21st-century river and land management

Location:	NY503129; N54°30.549 W002°46.141
Height:	440 metres
Geology:	Borrowdale Volcanic Group
Ownership:	United Utilities
Access:	Open access land; easily accessible from the Old Corpse Road
Website panorama:	From south-west through south to north-east

Thorny Knott (highest point on the skyline) viewed from the south

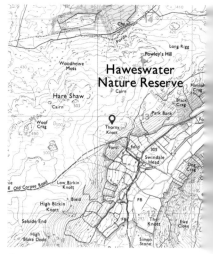

Location of Thorny Knott

♀ The View

There is nowhere else quite like this in the Lake District. Here is a remote valley that, over the centuries, has typically supported a dozen or so families following an agro-pastoral way of life and its traditions. Today, surrounded by bleak, open moorlands, the valley is quiet, barely populated by people or sheep and gives the appearance that nothing is going on.

However, this is United Utilities territory, managed by the Royal Society for the Protection of Birds, where innovative land- and river-management practices are being trialled. Here, perhaps, is a glimpse of how some upland Lake District landscapes might evolve in the future.

View from Thorny Knott looking east

Your viewing position: Looking south-east across the valley

Valley shapes and waterfalls: Below you and to your right there are contrasting valley shapes – the glacial trough of Swindale, its trough end at Dodd Bottom, the hanging valley of Mosedale and the prominent combe in the upper reaches of Hobgrumble Gill. The ice moving north-east off the high central fells during the Ice Age carved out the trough of Swindale, apparently influenced by the line of the Swindale Fault that runs east-north-east to west-south-west roughly along the line of the valley and the Old Corpse Road. Mosedale Beck and Hobgrumble Gill are hanging valleys cascading down 140 metres in height into the glaciated valley and then combining to create Swindale Beck. In contrast to Swindale, Mosedale is a shallow and broad V-shaped valley that is thought to have been subject to only light glaciation, a common characteristic of much of the eastern Lake District. The height of Thorny Knott where you are sitting is about the same as the floor of Mosedale, emphasising the differences in glacial erosion that have taken place between the two locations. However, there are some doubts as to whether Mosedale Force is solely the consequence of the glacial deepening of Swindale; it is possible that there was a marked drop in the longitudinal profile of the pre-glacial valley (known as a 'knick point') that was later exaggerated by glacial action.

A short-lived glacier: On the valley floor to your right are some prominent moraine ridges and mounds to the left of Dodd Bottom. These features are evidence of a small glacier that formed during the Younger Dryas, the short cold period that began 13,000 years ago. The glacier is thought to have filled much of the Dodd Bottom trough end (which some geomorphologists view as a corrie). In this north-east-facing position, wind-driven snow accumulated, formed glacier ice and further hollowed out the trough end, depositing rock debris as moraines. The grassy combe in upper Hobgrumble Gill appears to be a suitable location for snow and ice accumulation to feed the glacier but evidence of ice accumulation is lacking; nor is there evidence of ice movement feeding from the higher ground above. So the thinking is that the glacier was nourished solely within the confines of the valley and probably developed up to, but not above, Nabs and Geordie Greathead Crags, with its lower limit marked by the moraines.

Settlement history: The present scene of a quiet and barely populated valley is not what you would have seen through much of the medieval period or in later centuries. The earliest evidence of human presence in the area is from prehistoric funerary monuments, such as round and long cairns, and agricultural land earthworks probably dating from the Bronze and Iron ages. There are a number of sites on the moorland across the valley from where you are sitting, at White Raise, Waite Howes and to your right up on Selside Pike, some of which are protected ancient monuments. By the medieval period, the valley itself was populated with a number of farmsteads, probably linked in a number of ways with Shap Abbey, which is known to have acquired pasture in the valley in 1191. By the end of the medieval period most of the useful land in the valley bottom was enclosed and there were only small additional intakes in the 16th and 17th centuries. In the 19th century all the land in your view, outside the already enclosed fields, avoided Parliamentary Enclosure and remains unenclosed common land today This is in contrast to the tracts of land south of a line from Selside Pike to Wet Sleddale where long straight Parliamentary Enclosure walls are prominent in the landscape and can be seen on the Ordnance Survey map.

A chapel and school were built in the early 18th century at Truss Gap, down to your left, and the parsonage was built on the valley floor opposite you. The chapel was a "chapel-of-ease" to the parish church at Shap. The Old Corpse Road, which is close to where you are sitting, was the route by which corpses from Mardale were brought up over the fell and carried through Swindale for burial at Shap. During the latter part of the 19th century the farming fortunes of the valley declined; the population fell from 45 inhabitants in 1850 to 13 in 1900. The farmsteads at High Swindale Head and Swindale Foot and other farms in the valley were abandoned and in due course the chapel, school, parsonage and other buildings disappeared.

Swindale School

United Utilities ownership: All the land in your view is owned by the water company United Utilities as part of their Haweswater estate. United Utilities is a successor to Manchester Corporation, which in 1919 acquired Swindale Beck and all the land within its catchment, together with all of Mardale and Wet Sleddale, with the intention to build reservoirs to meet the need for additional public water supplies in the expanding city. The proposals raised considerable public objection at the time and the plan for Swindale did not proceed although Manchester Corporation had got as far as demolishing the chapel in 1938 in anticipation of work starting. Instead of a reservoir it was decided to extract water from Swindale Beck and transport it to the Haweswater reservoir via an underground aqueduct that was finally commissioned in 1957. The point where the water is extracted is at the weir a little downstream of Truss Gap down to your left. Using the Swindale Beck catchment as a gathering area for public drinking water and for sheep grazing created a number challenges, notably soil erosion and sediment transport to water abstraction points and the risks associated with cryptosporidium getting into the water supply via animal dung. Therefore, in 2012 United Utilities awarded the tenancies for the farms it owns in Swindale and in the Naddle valley, to the north, to the RSPB with a view to trialling new sustainable upland farming practices that could meet the objectives of both organisations (see pages on Brown Howe, Mardale).

Swindale Beck restoration: Since 2015, as a part of this approach by United Utilities and the RSPB, Swindale Beck and its catchment have been the scene for some highly innovative water- and land-management interventions undertaken with support from the Environment Agency and Natural England. The valley's natural habitats were not in a good ecological condition. This prompted the development of a Swindale Beck restoration project to restore the natural river and ecological processes in expectation that this would enhance biodiversity, alleviate flooding, capture carbon and reduce the cost of treating public water supplies.

For at least 160 years, Swindale Beck was in an artificially straightened and rock-armoured channel cut off from its flood plain. This created a number of problems: floodwater that over-topped the artificial raised banks could not return to the river and would instead create stagnant pools in the meadowland; within the straightened channel there were few slower-flowing areas where suitable habitats for fish spawning could form; and the fast flow in the channel carried gravel and silt to the public-water-supply extraction point just downstream of Truss Gap. In response, a scheme was implemented to return the river regime to its original meandering course. From where you are sitting you can probably see that the line followed by the beck is not the same as on your older Ordnance Survey map. A new sinuous alignment for the beck was designed, based on old river channels. A channel with a simple profile was dug by contractors during 2016 and natural processes did the rest with surprising speed, creating 890 metres of restored and new meandering channel with new gravel bars, shallow riffles and pools from Truss Gap upstream almost to Dry Grove Gill. The old river channel was filled in and sown with seed from an adjacent hay meadow, extending this botanically rich area. For a further 430 metres upstream, a more limited intervention removed rock armour and placed some removed rocks in the channel to create riffles. The whole beck is now morphologically more diverse and the evidence is that ecological diversity is now returning. Around 160 years ago, the local community took on the task of straightening the beck to create useable meadows and allow more productive farming. Very deliberately undoing that hard work can seem almost disrespectful, but it can also be seen as an example of how a cultural landscape changes as successive land owners and occupiers strive to meet the priorities of their own time, as they see them.

The shallow valley of Mosedale situated within an extensive landscape of blanket bog; Mosedale Cottage to the left

Other environmental work: Complementing this work, just downstream of Truss Gap where water is taken into the underground aqueduct to Haweswater, the weir and associated facilities have been upgraded to improve ecological conditions downstream, improve the quality of water being taken into the aqueduct and to allow fish migration upstream through fish and eel passes. Additionally, within the wider catchment various projects have been implemented to reduce flooding, following the "slow the flow" principle, and to assist carbon retention and sequestration (capturing carbon from the atmosphere). Peat restoration work has taken place in Mosedale to enable the Lake District's largest area of blanket bog to recover: 47 kilometres of artificial drain have been blocked and all livestock removed for 10 years. Other areas of bogs and mires have received similar treatment, including the area just beyond and to the left of The Knott on the valley floor to your right. In the catchment as a whole, some 40,000 trees have been planted, including 3,000 on the land in front of The Knott that will also serve as a flood attenuation area.

Acknowledgements

Firstly, I give my thanks to all those who have written about the Lake District before me and published in print and on-line. The combined resource they have created is huge and provided me with the very wide range of information and understanding of events that this guide required. In particular I feel I must highlight the four-volume nomination dossier that was prepared by the Lake District National Park Partnership and submitted to UNESCO in the successful bid for World Heritage status for the English Lake District. It is a really stunning piece of work that I heartily recommend for reading (a bit at a time though, it's very long!). When I found that books or experts were not in agreement, it was to this dossier that I often turned for guidance.

I am indebted to those individuals who gave of their time to read over sections of the guide in draft and offer comments and suggestions: Warren Allison, Jeff Cowton, Joe Daniels, Derek Denman, Elaine Essery, Peter Ford, Anthea Jones, Jamie Lund, Rachel Oakley, Lee Schofield, Alan Smith, Gareth Thornley, Diana Whaley and Sylvia Woodhead. I am grateful for their contributions, but of course any mistakes are solely my responsibility.

My thanks to the Lake District National Park Authority for their support, in particular the help and guidance provided by Stephen Ratcliffe and Mairi Lock, and to Alex McCoskrie at the National Trust who sourced many photographs and assisted in other ways.

Also thanks to Graham Barron who took on the task of developing the guide's website and to Chris Bagshaw who at an early stage gave me a cautionary tutorial about the foreign country that is book publishing.

The good people at Libri Publishing have been a joy to work with. Their guidance and patience with this novice author has made the publication process largely stress free, at least for me. I am grateful for their boldness in taking on this project.

Finally, thanks to my wife Valerie for her unstinting support while this guide slowly took shape. Over the years she had got used to me disappearing to go climbing; then she had to get used to me disappearing to research and write. Her understanding and forbearance have been heroic.

Glossary

Agro-pastoral. The practice of agriculture that includes both the growing of crops and raising livestock.

Ancient monument. A building or site entered on a national schedule by virtue of displaying heritage interest of national importance.

Ancient woodland. Land that has been continuously wooded since at least 1600. Ancient woodland includes "ancient semi-natural woodland" composed of native tree species not obviously planted, and also "plantations on ancient woodland sites" where the site, although replanted by non-native trees, retains many ancient woodland features and can respond well to restoration.

Barony. A form of land tenure held by Norman barons in return for service to the king.

Beck. Small river or stream.

Bloomery. A medieval charcoal-fired furnace that smelted iron ore to produce a lump of iron and slag, the "bloom", which was further worked to produce wrought iron.

Borrowdale Volcanic Group. A group of rocks formed during a violent volcanic period 450 million years ago during the Ordovician period.

Bronze Age. The period between 2000 and 800 BC characterised initially by the use of copper and then by bronze made by combining copper and tin.

Caledonian Orogeny. A mountain-building period 400 million years ago, after the Silurian period.

Carbon management. In the context of rural land management, carbon management is particularly about taking actions to maximise the storage of carbon in soils (especially peat) and vegetation (especially woodland).

Col. The lowest point on a mountain ridge between two peaks.

Common field. Typically an area of the valley floor for the growing of crops and grazing of livestock that was managed and used according to communal practice and custom. Sometimes referred to as "open field".

Common land. Land that has an owner but over which other people have various rights, such as to pasture animals, cut wood, quarry for stone or cut peat. Mainly fell land, much of which was designated "forest" during the medieval period. Sometimes referred to as "waste".

Continuous cover forestry. A management approach that works with the particular characteristics of a forest site, maintaining a permanent irregular structure through selection and harvesting of individual trees. Generally, clear felling is avoided.

Coppice. Woodland that has been managed by regular cutting every 15 years or so to produce a crop of straight poles from a single base. Used for a variety of purposes, including for making charcoal.

Corrie. The commonly used name for large armchair-shaped fellside hollows excavated by ice. Called "combes" and "coves" in the Lake District and "cirques" by geomorphologists.

Customary tenure. Land held securely as a tenant with ancient customary rights, including the right to pass land down to sons, and subject to paying dues to the lord of the manor.

Demesne. Land that was managed directly by the lord of the manor for his own needs.

Devensian. The last glacial period that began 118,000 years ago. A period of fluctuating temperatures with a last glacial maximum 20,000 years ago. See *Ice Age*.

Drumlin. Smooth oval or elliptical hills composed of glacial till or water-sorted sands and gravels and aligned with the direction of ice movement. Formed beneath glaciers and ice sheets.

Fell. High ground. Used generally to refer to the hills and mountains of the Lake District.

Forest. A Norman baron's hunting ground subject to forest law to protect game.

Gill (or Ghyll). Small stream, often flowing steeply with waterfalls and cascades.

Glacial till. Unsorted eroded material deposited by a glacier.

Glaciofluvial deposits. Sediments carried and laid down by water issuing from ice sheets and glaciers.

Grange. A farm run by one of the landowning monastic houses.

Heaf (or heft). An area of open unenclosed fell that lambs learn from their mothers as "their" patch. Enables different farmers and their flocks to share open grazing landscapes.

Hercynian Orogeny. A mountain-building period about 300 million years ago during which the rocks of the Lake District were uplifted into a gentle dome.

Herdwick. An upland breed of sheep native to Cumbria. Ancestors of the breed are thought to have been introduced by the Norse.

Ice Age. In this book the term refers to the most recent glacial episode, which occurred during the late Devensian period. It began around 33,000 years ago, reached its maximum extent 20,000 years ago and was over by 15,000 years ago. There were many Ice Ages before and during the Devensian period, but their effect on the present landscape is largely unknown. Also see *Younger Dryas.*

Igneous. Rocks formed from the cooling of molten magma. Can be extruded onto the earth's surface, such as lava, or intruded into the earth's crust, such as granites.

Inbye. Enclosed land of a farm on the valley floor.

Intake. An area of land enclosed from the common between the inbye and the open fell.

Iron Age. The period between 800 BC and 100 AD.

Listed Building. A building listed on the National Heritage List that is of special architectural or historic interest of national importance. There are three categories of significance – grades I, II* and II.

Manor. A medieval estate held by a lord of the manor who had the right to establish a manor court to adjudicate on disputes.

Medieval. The period from the Norman Conquest to 1600.

Mesolithic. The period 8000 to 4000 BC when larger groups of hunter gatherers came into the Lake District.

Metamorphic. Igneous and sedimentary rocks that have been altered by heat or pressure, such as slate.

Moraine. Ridges and mounds of glacial till.

Neolithic. The period 4000 to 2000 BC during which arable agriculture developed.

Open field. See *common field*.

Ordovician. The geological period 490 to 440 million years ago during which the Skiddaw Group and Borrowdale Volcanic Group rocks were laid down.

Park. An enclosed area usually referring to a deer park, some of which evolved into ornamental parkland.

Parliamentary Enclosure. A series of Acts of Parliament that led to the enclosure of open fields and common land creating private property rights to land previously held in common. Also brought about much associated reorganisation of land use and management.

Picturesque. A movement in the late 18th century that arose as a reaction to classical forms in architecture and art. In a landscape context it embraced irregularity of form and the intermingling of visual stimuli within certain principles of composition, usually with a rustic setting.

Pinfold. An enclosure for temporarily holding stray animals.

Pitstead. A platform for the manufacture of charcoal, usually from coppice wood.

Plantation on ancient woodland site. See *ancient woodland*.

Priority habitats. Habitats of principal importance for the conservation of biodiversity as listed under the provisions of section 41 of the Natural Environment and Rural Communities Act 2006.

Pyroclastic. Rocks derived from material blown violently into the atmosphere by volcanic explosions.

Ring garth. A stone wall or fence that divided the cultivated valley-bottom fields from grazed fell land.

Roche moutonnée. A glaciated asymmetric rock landform. Commonly interpreted from the French simply as "sheep-like rock". However, the term was first coined by the French Alpine explorer Horace-Benedict de Saussure in 1786. The shape of the rock reminded him of the wigs fashionable with the gentry at the time, which were smoothed over with mutton fat (hence moutonnée) to keep them in shape.

Romano-British. A culture that was a fusion of imported Roman culture with that of indigenous Britons during the period 100 to 400 AD.

Sedimentary. Rocks formed from the deposition of material in water or on land, such as mudstones, shales, limestone and sandstone.

Shieling. A temporary building used in the summer months usually at a valley head or above a valley. Associated with the management of stock.

Silurian. The geological period 440 to 415 million years ago when the Windermere Supergroup rocks were laid down.

Site of Special Scientific Interest. Land formally notified as being of special biological or geological interest in a national context.

Skiddaw Group. A group of marine sedimentary rocks formed during the early Ordovician period. The oldest rocks in the Lake District.

Special Area of Conservation. An area identified for the protection of habitats and species of European interest listed in the EU Habitats Directive.

Statesman. A name given to yeoman farmers in the Lake District who held their land by customary tenure; derived from "estatesmen".

Talus. Rockfall from crags, commonly referred to as "scree".

Tuff. A pyroclastic rock made of volcanic ash violently ejected from a vent during a volcanic eruption and then deposited and compressed.

UNESCO. United Nations Educational, Scientific and Cultural Organisation.

Vaccary. Large cattle/dairy farms usually under monastic control during the medieval period.

Viewing station. A location for viewing the landscape, particularly Picturesque landscapes in the late 18th and early 19th centuries. Popular with tourists who would turn their backs to the view and then hold up a tinted mirror (a Claude glass) that framed the view and assisted in drawing it.

Villa. In the 18th and 19th centuries in the Lake District, refers to a large residence with designed gardens, usually in an attractive lakeside setting.

Volcaniclastic. Rocks of volcanic origin moved and redeposited as sedimentary material, usually by water.

Wainwright. Alfred Wainwright was the writer and illustrator of a series of popular guides to the Lake District fells. He described 214 fells and their summits, sometimes referred to as "Wainwrights".

Waste. See *common land.*

Water Framework Directive. A European Directive adopted in 2000 for managing and protecting water bodies.

Windermere Supergroup. A group of marine sedimentary rocks formed during the Silurian period.

Younger Dryas. A cold period between 13,000 and 11,700 years ago when small glaciers reformed.

List of Hows and Knotts

Below is a list of all hows and knotts within the World Heritage Site named on the Ordnance Survey 1:25,000 map that are landforms and not named habitations. The boundary of the World Heritage Site is the boundary of the National Park prior to its extension in 2016. The coordinates for the kilometre square are at the south-west corner of the square.

Name		Km sq	Parish	Height (m)	Notes
Hows					
Armaside	How	NY1427	Lorton	91	
Ashness	How	NY1304	Wasdale	80	
Bald	Howe	NY4023	Matterdale	280	
Banking	Hows	SD2195	Dunnerdale	170	
Bass	How	SD3897	Claife	40	No protruding landform
Bellart	How	SD4484	Witherslack	20	
Bent	Haw	SD2392	Dunnerdale	200	
Benty	Howe	NY4613	Bampton	440	
Berry	How	NY1104	Wasdale	90	
Birch	How	NY2509	Borrowdale	300	
Birk	Haw	SD2692	Torver	160	
Birk	Haw	SD2690	Kirkby Ireleth	100	
Birk	How	NY3923	Matterdale	270	
Blake	How	SD2899	Coniston	600	
Bleaberry	Haws	SD2694	Torver	350	
Bleaberry	How	NY1904	Eskdale	410	
Bleak	Haw	SD1993	Dunnerdale	150	
Bleak	How	NY2712	Borrowdale	320	

Boat	How	NY1703	Eskdale	337	
Boat	How	NY1911	Ennerdale	500	
Boat	How	NY0810	Ennerdale	400	No protruding landform
Boat	How	NY1113	Ennerdale	363	View described in this guide
Boat	Howe	NY4711	Shap Rural	400	
Booth	How	SD2897	Coniston	390	
Bracken	How	NY3921	Matterdale	360	
Bracken	Howe	NY5313	Shap Rural	330	
Brackenthwaite	Hows	NY1521	Buttermere	208	
Braithwaite	How	NY2223	Above Derwent	172	
Brim Fell	Haws	SD2798	Coniston	600	
Broad	Haws	SD2496	Dunnerdale	250	
Broad	How	NY4208	Lakes	500	
Brock	How	SD2699	Dunnerdale	450	
Brock	How	SD2497	Dunnerdale	350	
Brown	Haw	SD2293	Dunnerdale	350	
Brown	How	NY3004	Lakes	469	
Brown	How	NY1406	Wasdale	350	No protruding landform
Brown	How	NY1115	Ennerdale	320	
Brown	Howe	NY2604	Lakes	380	
Brown	Howe	NY4812	Shap Rural	500	View described in this guide
Brown	Howe	NY4608	Kentmere	710	
Brown	Howe	NY5108	Shap Rural	550	
Brown	Howe	NY1925	Lorton	517	
Brown	How	NY3217	St Johns	350	
Brunt	How	NY3503	Skelwith	100	
Buckstone	Hows	NY2114	Buttermere	600	
Bull	How	NY2002	Eskdale	300	
Butharlyp	How	NY3307	Lakes	106	
Capple	Howe	NY4302	Windermere	445	
Carble	Hows	NY2600	Dunnerdale	630	
Carr	Howes	NY3303	Skelwith	80	
Castle	How	NY2534	Ireby	305	

Castle	How	NY2030	Wythop	90	
Castle	How	NY2300	Dunnerdale	273	View described in this guide
Castle	How	SD1892	Ulpha	144	View described in this guide
Castle	How	NY2903	Lakes	145	
Cat	How	NY1104	Wasdale	80	
Cawker	How	SD4089	Cartmel Fell	150	
Churn	How	NY2305	Eskdale	450	
Cockley	How	NY2417	Borrowdale	300	
Cockley	How	NY3913	Patterdale	300	
Coin	How	SD3192	Satterthwaite	220	
Copthwaite	How	NY2501	Dunnerdale	450	
Crab	Haws	SD2990	Colton	90	
Crowberry	Haws	SD2898	Coniston	310	
Dale	Hows	NY2637	Ireby	300	
Dale	How	NY1618	Buttermere	200	
Darling	How	NY1924	Lorton	480	
Darling	How	NY1928	Wythop	200	
Deer	Hows	NY3604	Lakes	380	
Deergarth	How	NY3116	St Johns		Island in Thirlmere
Dog	How	SD2196	Ulpha	290	
Ees	Hows	SD3695	Hawkshead	70	
Elf	Howe	NY4600	Over Staveley	220	
Eller	How	NY1803	Eskdale	280	No protruding landform
Eller	How	SD4181	Upper Allithwaite	178	View described in this guide
Ewe	How	NY2724	Underskiddaw	180	
Fair	How	NY3821	Matterdale	350	No protruding landform
Fisherty	How	NY3701	Claife	50	
Forked	How	SD2499	Dunnerdale	450	
Foss	How	SD2498	Dunnerdale	300	
Fox	Haw	SD2293	Dunnerdale	385	
Gaze Stone	How	NY2501	Dunnerdale	395	
Glade	Haw	SD2088	Broughton West	120	

Glade	How	NY1306	Wasdale	430	
Glede	Howe	NY5212	Shap Rural	476	
Great	How	NY3202	Skelwith	211	
Great	How	NY3118	St Johns	320	
Great	How	NY1904	Eskdale	522	
Great	Howe	NY4806	Longsleddale	450	
Great Arming	How	SD1899	Eskdale	260	
Great Castle	How	NY3007	Lakes	470	
Great Green	Hows	SD3589	Colton	229	
Great Round	How	NY2012	Buttermere	554	A grade 2/3 scramble
Green	How	SD1898	Eskdale	270	No protruding landform
Green	How	NY1906	Wasdale	540	No protruding landform
Green	How	SD2095	Ulpha	280	
Green	How	NY2537	Ireby	320	
Green	How	SD2792	Torver	185	
Green	How	NY2400	Dunnerdale	480	
Green	How	NY2500	Dunnerdale	600	
Green	How	SD1799	Eskdale	200	
Green	How	NY3821	Matterdale	340	No protruding landform
Griddle	How	NY3201	Skelwith	190	
Gummer's	How	SD3988	Staveley in C	321	View described in this guide
Hanging	How	NY0901	Irton	50	
Hart	How	NY2406	Eskdale	700	
Hart	Howe	SD4393	Crook	150	
Hartsop above	How	NY3812	Patterdale	586	A Wainwright summit
Hawes	How	NY3115	St Johns		An island in Thirlmere
Haws		SD2393	Dunnerdale	280	
Hawthwaite	Haw	SD2289	Kirkby Ireleth	120	
High	How	NY1625	Lorton	200	
High	Howe	NY4617	Bampton	550	No protruding landform
High	Hows	NY0920	Lamplugh	313	
High Blind	How	SD3897	Claife	270	
High Buck	How	NY2513	Borrowdale	450	View described in this guide

High Light	Haw	SD3090	Colton	263	
High Pike	Haw	SD2694	Torver	354	
High Spying	How	NY3514	Patterdale	860	
High Teighton	How	NY2703	Dunnerdale	500	
High Wether	Howe	NY5110	Shap Rural	531	
Hodge	How	NY1800	Eskdale	101	
Hodgson	How	NY2423	Above Derwent	99	
Hollin	How	NY3921	Matterdale	330	No protruding landform
Hollin	How	SD2297	Dunnerdale	270	
Hollin	Howe	NY4701	Over Staveley	280	
Hollin House	Haw	SD2396	Dunnerdale	180	
Horse	How	SD2597	Dunnerdale	500	
How	Top	NY3728	Mungrisdale	277	
How		NY2424	Above Derwent	94	
How		NY2814	Borrowdale	450	
How Scale	Haw	SD2295	Dunnerdale	200	
Howes		NY4910	Shap Rural	580	
Hutching's	How	NY1908	Wasdale	250	View described in this guide
Jopplety	How	NY2616	Borrowdale	420	
Kidbeck	How	NY1104	Wasdale	110	
Kids	Howe	NY5402	Fawcett Forest	250	No protruding landform
Kidson	How	SD2699	Dunnerdale	600	
Kidsty	Howes	NY4612	Bampton	510	
Kilbert	How	NY4018	Martindale	370	
Kiln	How	NY2327	Underskiddaw	80	
King's	How	NY2516	Borrowdale	392	
Kinney	How	NY0713	Ennerdale	360	No protruding landform
Kirk	Howe	NY2906	Lakes	120	
Kitt	How	SD1696	Ulpha	290	
Lad	How	SD2196	Ulpha	210	
Lad	Hows	NY1719	Buttermere	426	
Lang	How	NY3107	Lakes	414	
Leads	Howe	NY4408	Kentmere	450	

Limestone	Haws	SD2796	Coniston	400	
Ling	How	NY2429	Bassenthwaite	360	View described in this guide
Little	How	NY3119	St Johns	200	
Little Arming	How	SD1899	Eskdale	244	
Little Castle	How	NY3107	Lakes	490	
Little Green	Hows	SD3690	Colton	170	No protruding landform
Little Round	How	NY2013	Buttermere	494	
Loft Rigg	How	SD2499	Dunnerdale	450	
Long	Haws	SD2896	Coniston	250	
Long	How	NY1717	Buttermere	120	
Long	How	NY2601	Dunnerdale	600	
Looking	How	SD2499	Dunnerdale	400	
Looking	Howe	NY3008	Lakes	400	
Looking	How	SD4090	Cartmel Fell	164	
Lord's	How	NY2813	Borrowdale	490	
Low	How	NY2309	Borrowdale	550	
Low	How	NY3721	Matterdale	497	
Low Blind	How	SD3897	Claife	220	
Low Buck	How	NY2513	Borrowdale	240	
Low Light	Haw	SD3090	Colton	240	
Low Pike	Haw	SD2694	Torver	300	
Low Spying	How	NY3515	Patterdale	800	
Low Teighton	How	NY2702	Lakes	420	
Lower	Hows	NY2800	Coniston	650	
Lunsty	How	NY4503	Kentmere	150	
Mickle	How	NY3238	Caldbeck	240	
Middle	How	NY2911	St Johns	483	
Middle	How	SD3792	Satterthwaite	140	
Moor	How	SD3991	Cartmel Fell	229	
Moor	How	SD2899	Coniston	350	
Moor	Howe	NY4200	Windermere	210	
Naithwaite	How	NY1830	Embleton	80	
Nether	How	NY1716	Buttermere	120	

Nitting	Haws	NY2416	Borrowdale	360	A grade 1 scramble
Oak	Howe	NY3005	Lakes	100	
Pattison	How	SD4085	Allithwaite Upper	177	
Pike	How	NY4108	Lakes	620	
Pike	How	SD1697	Eskdale	260	
Pike	How	SD2399	Dunnerdale	210	
Pike	Howe	NY2806	Lakes	400	View described in this guide
Piked	How	SD2897	Coniston	410	
Piked	Howes	NY4404	Kentmere	450	
Pinnacle	Howe	NY4916	Bampton	380	
Pow	How	NY2422	Above Derwent	100	No protruding landform
Powter	How	NY2226	Lorton	230	
Rabbit	How	SD1399	Eskdale	73	
Rabbit Cat	How	SD0796	Drigg	5	No protruding landform
Rake	How	NY3122	St Johns	120	
Rascal	How	SD2998	Coniston	300	View described in this guide; a grade 2 scramble
Rasp	Howe	NY4603	Kentmere	380	View described in this guide
Ratten	Haw	SD2690	Kirkby Ireleth	180	
Raven	Howe	NY4514	Martindale	700	No protruding landform
Raven Nest	How	SD2599	Dunnerdale	400	A grade 2 scramble
Red	How	NY2502	Dunnerdale	600	A grade 1 scramble
Rough	How	NY1601	Eskdale	240	
Rough	How	SD1796	Ulpha	260	No protruding landform
Rough	Hows	SD3798	Claife	190	
Round	Haw	SD2788	Blawith	200	
Round	How	NY4016	Martindale	620	
Round	How	NY2108	Wasdale	741	
Round	How	NY4111	Patterdale	618	
Round	How	NY3920	Matterdale	387	
Round	How	NY2725	Underskiddaw	200	
Round	How	NY2601	Dunnerdale	650	

Rowan Tree	How	NY2500	Dunnerdale	450	
Rowantree	How	SD1595	Ulpha	400	
Rowantree	How	SD2297	Dunnerdale	270	
Saddlebacked	How	NY2300	Dunnerdale	250	
Sale	How	NY2728	Underskiddaw	666	
Scott	Howe	SD4292	Crook	169	
Seat	How	NY2125	Above Derwent	490	
Seat	How	SD1697	Eskdale	311	
Seathwaite	How	NY1730	Embleton	190	
Shooting	How	NY1408	Wasdale	580	
Shudderstone	How	SD2598	Dunnerdale	510	A grade 2 scramble
Silver	How	NY3206	Lakes	395	View described in this guide; a Wainwright
Silver	How	SD1998	Eskdale	320	
Sleet	How	NY2022	Above Derwent	550	
Snipes	How	NY2921	St Johns	240	
Sour	Howes	NY4203	Windermere	483	A Wainwright summit
Spying	How	NY3819	Matterdale	350	No protruding landform
Stang	How	NY2113	Buttermere	320	
Stanger	How	NY1427	Embleton	79	
Stephen	How	NY3302	Skelwith	250	
Stephenson	Haw	SD2293	Ulpha	320	
Stone	How	NY5414	Shap	260	No protruding landform
Stone	How	NY5212	Shap Rural	400	
Stone Guard	How	SD2599	Dunnerdale	600	
Stunfell	Howe	NY4404	Kentmere	400	
Swanesty	How	NY2615	Borrowdale	300	
Swanesty	How	NY2417	Borrowdale	160	
Swinsty	How	SD2398	Dunnerdale	230	
Swirl	How	NY2700	Coniston	802	A Wainwright summit
Tarn	Hows	SD3299	Coniston	200	
Tewit	How	NY1411	Ennerdale	610	
The	Howe	SD4588	Crosthwaite	72	

The	Howes	NY5018	Bampton	260	
The	Hows	SD2399	Dunnerdale	220	
The	Haws	SD1994	Ulpha	250	
The	How	NY2514	Borrowdale	100	
Thorn	How	SD1997	Eskdale	360	
Toad	How	NY2109	Wasdale	450	
Tongue	How	NY0709	Ennerdale	240	No protruding landform
Vicker's	How	NY1605	Wasdale	300	
Waite	Howes	NY5112	Shap Rural	380	
Wether	How	SD2599	Dunnerdale	740	
Wether	How	NY2601	Dunnerdale	700	
Whelpsty	How	SD1791	Ulpha	230	
Whinny	Howe	NY4101	Windermere	240	
Whirl	Howe	NY4905	Longsleddale	180	
White	How	SD2097	Dunnerdale	444	
White	How	SD1690	Millom Without	350	
White	How	SD2499	Dunnerdale	350	
White	Howe	NY3807	Lakes	350	
White	Howe	NY5204	Fawcett Forest	530	
Wilkes	How	NY1123	Blindbothel	263	No protruding landform
Windy	Howe	SD4586	Crosthwaite	90	
Winscale	Hows	NY1408	Wasdale	500	
Withered	Howe	NY4606	Kentmere	400	
Wolf	Howe	NY5402	Fawcett Forest	331	
Wood	Howe	NY4711	Shap Rural		Island in Haweswater
Work	How	NY3905	Lakes	350	
Worm	How	SD2397	Dunnerdale	200	
Wormshell	How	SD2097	Dunnerdale	420	
Yokerill	Hows	NY1107	Wasdale	280	No protruding landform

Knotts

Aikin	Knott	NY2119	Above Derwent	475	
Allen	Knott	NY4101	Windermere	220	View described in this guide

Ash	Knott	NY5010	Shap Rural	480	
Askill	Knott	NY1222	Loweswater	284	
Band	Knotts	NY4508	Kentmere	600	
Bedafell	Knott	NY4216	Martindale	480	
Bell	Knott	SD2689	Kirkby Ireleth	200	
Bell	Knott	NY3910	Patterdale	470	View described in this guide
Bell	Knott	NY3819	Matterdale	350	View described in this guide
Belles	Knott	NY2908	Lakes	480	A grade 2 scramble
Belt	Knott	NY2615	Borrowdale	200	
Belt	Knott	NY2612	Borrowdale	300	
Bennesty	Knott	NY3222	St Johns	600	
Birk	Knott	SD2990	Colton	120	
Birk	Knott	NY2904	Lakes	330	
Birkie	Knott	NY4318	Martindale	350	
Black	Knott	NY2912	St Johns	660	
Black	Knott	NY2714	Borrowdale	460	
Blackbeck	Knotts	NY1509	Wasdale	470	
Blawith	Knott	SD2688	Blawith	248	
Bleaberry	Knott	NY4018	Martindale	480	
Bleaberry	Knott	NY2804	Lakes	520	
Bleaberry	Knott	NY3912	Patterdale	512	
Bleak	Knott	SD2790	Blawith	200	
Bowder	Knott	SD2889	Blawith	130	
Bowness	Knott	NY1115	Ennerdale	333	
Brackenwife	Knotts	NY3409	Lakes	400	
Brimming	Knott	NY2815	Borrowdale	400	
Brown	Knotts	NY2719	Borrowdale	350	View described in this guide
Brown	Knott	NY2530	Bassenthwaite	410	
Brunt	Knott	NY3328	Mungrisdale	600	
Brunt	Knott	NY4800	Over Staveley	427	
Bursting	Knott	NY2109	Wasdale	350	
Buzzard	Knott	NY2529	Bassenthwaite	500	
Carling	Knott	NY1220	Loweswater	519	

Coldkeld	Knotts	NY2207	Eskdale	800	
Cop	Knott	NY2611	Borrowdale	460	
Cowsty	Knotts	NY4504	Kentmere	400	
Criscliffe	Knotts	NY2108	Wasdale	650	
Dod	Knott	NY2100	Eskdale	300	View described in this guide
Ether	Knott	NY2617	Borrowdale	410	
Fairy	Knott	NY3830	Mungrisdale	280	
Flass	Knotts	NY2108	Wasdale	500	
Force	Knott	SD1686	Whicham	140	
Force	Knott	SD3491	Satterthwaite	160	
Gibson	Knott	NY3110	Lakes	422	A Wainwright summit
Gladstone	Knott	NY2504	Lakes	700	
Glass	Knott	SD3289	Colton	78	
Great	Knott	NY2504	Lakes	696	
Great	Knott	SD3391	Satterthwaite	150	
Great	Knott	NY2429	Bassenthwaite	445	
Great	Knott	NY1609	Wasdale	500	
Greenhow	Knott	NY2310	Borrowdale	310	
Grey	Knotts	NY2112	Buttermere	697	A Wainwright summit; a grade 1 scramble
Gunson	Knott	NY2505	Lakes	822	
Hanging	Knotts	NY2407	Borrowdale	650	
Hard	Knott	NY2302	Dunnerdale	549	A Wainwright summit
Hay	Knott	NY3036	Caldbeck	400	
High	Knott	NY4500	Staveley	260	
High	Knott	NY2512	Borrowdale	500	
High	Knott	NY4118	Martindale	350	
High Birkin	Knott	NY4912	Shap Rural	550	
High Bleaberry	Knott	NY2505	Lakes	550	
High Buzzard	Knott	NY2509	Borrowdale	450	
High Great	Knott	NY4107	Lakes	380	
High Steel	Knott	NY2416	Borrowdale	450	
High Wax	Knott	NY1814	Buttermere	380	

Holmgill	Knotts	NY1909	Wasdale	450	
Hook	Knott	SD1685	Whicham	170	
Hovel	Knott	SD2191	Dunnerdale	220	
Inking	Knott	SD2792	Torver	150	
Ivy	Knott	NY2611	Borrowdale	300	
Kern	Knotts	NY2109	Wasdale	520	
Knott	Halloo	NY3126	Threlkeld	650	
Knott		NY2932	Mungrisdale	710	A Wainwright summit
Knotts		NY3001	Lakes	234	
Knotts		NY3922	Matterdale	390	
Knotts		SD2699	Dunnerdale	550	
Knotts		NY3427	Mungrisdale	420	
Knotts		NY4820	Askham	406	
Knotts		NY2025	Lorton	490	
Knotts		NY2614	Borrowdale	400	
Knotts of the	Tongue	NY2307	Eskdale	750	
Knotts		NY2025	Lorton	480	
Lad	Knott	NY3220	St Johns	350	
Lambing	Knott	NY1915	Buttermere	200	View described in this guide
Lamper	Knott	NY2611	Borrowdale	280	
Lease	Knott	SD1296	Muncaster	130	
Little	Knott	NY2430	Bassenthwaite	350	
Little	Knott	SD3392	Satterthwaite	190	
Loup	Knott	NY4621	Barton	470	View described in this guide
Low Birkin	Knott	NY4912	Shap Rural	500	
Low Bleaberry	Knott	NY2505	Lakes	450	
Low Buzzard	Knott	NY2509	Borrowdale	400	
Low Great	Knott	NY4106	Lakes	320	
Low Wax	Knott	NY1814	Buttermere	250	
Lower Kern	Knotts	NY2109	Wasdale	470	
Mart	Knott	NY1213	Ennerdale	320	
Middleboot	Knotts	NY2108	Wasdale	650	
Millrigg	Knott	NY4601	Over Staveley	278	

Mitchell	Knotts	SD3895	Claife	184	
Murthwaite	Knott	NY5100	Whitwell	300	
Newtown	Knott	SD0995	Muncaster	80	View described in this guide
Pianet	Knott	NY2304	Eskdale	430	
Piketoe	Knott	NY3216	St Johns	400	
Proud	Knott	NY1812	Ennerdale	550	
Rannerdale	Knotts	NY1618	Buttermere	355	A Wainwright summit
Red	Knott	NY2217	Above Derwent	350	
Reggle	Knott	NY3312	St Johns	410	
Rough	Knott	NY2816	Borrowdale	400	
Round	Knott	NY3333	Mungrisdale	603	
Rowentree	Knotts	NY4405	Kentmere	400	
Saddler's	Knott	NY1018	Lamplugh	365	
Saletarn	Knotts	NY4405	Kentmere	470	
Sandhill	Knotts	NY4815	Bampton	330	
Scale	Knott	NY1517	Buttermere	250	
Scale	Knotts	NY4505	Kentmere	360	
Scalebarrow	Knott	NY5115	Shap Rural	330	
School	Knott	SD4297	Windermere	232	
Seavy	Knott	NY2012	Ennerdale	450	A grade 3 scramble
Sharp	Knott	NY1020	Lamplugh	482	
Shipman	Knotts	NY4706	Longsleddale	587	A Wainwright summit
Shivery	Knott	NY2815	Borrowdale	490	
Silly	Knott	SD3583	Haverthwaite	184	
Silver	Knott	SD1399	Eskdale	170	
Slate	Knott	NY2301	Dunnerdale	400	
Smallthwaite	Knott	NY4508	Kentmere	480	
Smooth	Knott	SD3292	Satterthwaite	245	
Squat	Knotts	NY2216	Above Derwent	550	
Stair	Knott	NY1213	Ennerdale	260	
Starebeck	Knotts	NY1207	Wasdale	380	
Steel	Knotts	NY4418	Martindale	420	A Wainwright summit
Sty	Knotts	SD1485	Whicham	250	

Summer House	Knott	SD3787	Colton	185	
Swainson	Knott	NY0708	St Bridget Beckermet	345	
Swine	Knott	NY2907	Lakes	300	
Swineside	Knott	NY3719	Matterdale	553	
Symonds	Knott	NY2006	Wasdale	959	
The	Knott	SD2493	Broughton West	332	
The	Knott	SD2291	Dunnerdale	284	
The	Knotts	SD2887	Blawith	90	
The	Knott	SD2590	Kirkby Ireleth	130	
The	Knotts	NY4321	Matterdale	274	
The	Knott	SD1495	Muncaster	331	
The	Knott	SD2196	Ulpha	300	
The	Knott	NY4312	Patterdale	739	A Wainwright summit
The	Knott	SD2187	Broughton	60	
The	Knott	NY5011	Shap Rural	297	
Thorn	Knott	NY1206	Wasdale	350	
Thorny	Knott	NY5012	Shap Rural	440	View described in this guide
Threlkeld	Knotts	NY3223	Threlkeld	500	
Thunacar	Knott	NY2708	Lakes	723	A Wainwright summit
Thwaitehill	Knotts	NY2125	Above Derwent	310	
Timley	Knott	SD2897	Coniston	300	
Townend	Knotts	SD1283	Whicham	310	
Trap	Knotts	NY2418	Above Derwent	375	
Waterside	Knott	SD3686	Colton	120	
Wet	Knotts	NY2114	Buttermere	400	
Whelter	Knotts	NY4713	Bampton	350	
Whinny	Knott	SD4089	Cartmel Fell	210	
White	Knotts	NY4519	Martindale	500	
White	Knott	NY4621	Barton	420	
White Hall	Knott	SD1585	Whicham	311	
Willy	Knott	NY2836	Caldbeck	370	

Wilson	Knott	SD3899	Claife	150
Windy	Knott	SD1586	Whicham	373
Wood	Knotts	SD1795	Ulpha	350
Wool	Knott	SD2789	Blawith	220
Yewdel	Knott	NY2418	Borrowdale	150

Photograph Credits

All photographs have been taken by the author, with the exception of the following:

Page No.	Credit
27	© National Trust Images/Jamie Lund
30	© National Trust Images/James Dobson
31	J. Hardman/Museum of Lakeland Life and Industry, Lakeland Arts, Kendal
36	© National Trust Images/Steven Barber
42	Rafael Garea-Balado/Alarmy Stock Photo
65	© National Trust Images/Arnhel de Serra
67	© English Lakes Hotels Resorts and Venues
79	© Patterdale Today
85	Courtesy Wild Ennerdale Project
89	CumbriaImageBank.org.uk
90	J. Hardman/Museum of Lakeland Life and Industry, Lakeland Arts, Kendal
94	Cameni Images/Alarmy Stock Photo
97	© National Trust Images
101	© Graham Barron
106	© Graham Barron
109	© John Hodgson
112	© C. Headley/LDNPA
121	J. Hardman/Museum of Lakeland Life and Industry, Lakeland Arts, Kendal
125	CumbriaImageBank.org.uk
127	CumbriaImageBank.org.uk
132	© National Trust Images/Jamie Lund
137	CumbriaImageBank.org.uk
139	By kind permission of the Fell and Rock Climbing Club of the English Lake District www.frcc.co.uk
151	© National Trust Images/Paul Harris
157	Ullswater Steamers/Flickr
161	CumbriaImageBank.org.uk
162	We acknowledge permission given to use this image given by the Cumberland and Westmorland Antiquarian and Archaeological Society
169	J. Hardman/Museum of Lakeland Life and Industry, Lakeland Arts, Kendal
174	© Peter Trimming (cc-by-sa/2.0)
179	© Antiquary/Creative Commons
181	CumbriaImageBank.org.uk (Joseph Hardman Collection)
191	CumbriaImageBank.org.uk

Maps on pages 21, 52, 62, 68, 74, 80, 86, 92, 98, 104, 110, 116, 122, 128, 134, 140, 146, 152, 158, 164, 170, 176, 182, 188 © Crown Copyright and Database Rights 2019 Ordnance Survey Licence 100061047

Select Index of Places

About the Author

Guy Richardson has lived in Cumbria for over 30 years, climbing and walking in the Lake District, across the UK and abroad. He read geography at Oxford University and then pursued a career in town planning until retiring as Head of Planning with Cumbria County Council. He was a member of the Lake District National Park Partnership for a number of years and is a Fellow of the Royal Geographical Society. He now lives near Grange-over-Sands with his wife Valerie.